Sustainable Development Goals Series

The **Sustainable Development Goals Series** is Springer Nature's inaugural cross-imprint book series that addresses and supports the United Nations' seventeen Sustainable Development Goals. The series fosters comprehensive research focused on these global targets and endeavours to address some of society's greatest grand challenges. The SDGs are inherently multidisciplinary, and they bring people working across different fields together and working towards a common goal. In this spirit, the Sustainable Development Goals series is the first at Springer Nature to publish books under both the Springer and Palgrave Macmillan imprints, bringing the strengths of our imprints together.

The Sustainable Development Goals Series is organized into eighteen subseries: one subseries based around each of the seventeen respective Sustainable Development Goals, and an eighteenth subseries, "Connecting the Goals," which serves as a home for volumes addressing multiple goals or studying the SDGs as a whole. Each subseries is guided by an expert Subseries Advisor with years or decades of experience studying and addressing core components of their respective Goal.

The SDG Series has a remit as broad as the SDGs themselves, and contributions are welcome from scientists, academics, policymakers, and researchers working in fields related to any of the seventeen goals. If you are interested in contributing a monograph or curated volume to the series, please contact the Publishers: Zachary Romano [Springer; zachary.romano@springer.com] and Rachael Ballard [Palgrave Macmillan; rachael.ballard@palgrave.com].

More information about this series at http://link.springer.com/series/15486

Haris Alibašić

Strategic Resilience and Sustainability Planning

Management Strategies for Sustainable and Climate-Resilient Communities and Organizations

Second Edition

Springer

Haris Alibašić
Public Administration Program
University of West Florida
Pensacola Beach, FL, USA

ISSN 2523-3084 ISSN 2523-3092 (electronic)
Sustainable Development Goals Series
ISBN 978-3-030-91158-4 ISBN 978-3-030-91159-1 (eBook)
https://doi.org/10.1007/978-3-030-91159-1

This Springer imprint is published by the registered company Springer Nature Switzerland AG
The registered company address is: Gewerbestrasse 11, 6330 Cham, Switzerland

God completed them as seven firmaments in two periods and assigned to each heaven order and function. And adorned the nearest heaven with stars, and provided it with guard. Such is the design of the Omnipotent, the Omniscient.

Quran: Chapter 41:12, Fussilat
(Explained in Detail)

Dedicated to my resilient family, my children Jakub, Lamija, Imana, and Harun, my parents Emira and Dževad, and my brothers Venso and Emir for their unconditional love and support. In memory of my late grandparents Fatima and Jakub Alibašić, and Fatima and Rahman Ibraković, who were early sustainability adopters and who taught me to love the Earth and respect nature and the environment.

Preface

The book examines and develops strategies for an effective strategic resilience and sustainability plan for communities and local government organizations, such as cities, counties, villages, and townships. It offers advanced methods for evaluating resilience and sustainability strategies in organizations and communities. As municipal governments and communities responded to the COVID-19 pandemic, strategic resilience and sustainability planning was instrumental in responding to the crisis. The recommendations in this book are based on extensive academic research and expertise in directing applied resilience and sustainability programs.

Diverse aspects of resilience from climate change, climate preparedness, readiness, Quadruple Bottom Line strategy, greenhouse gas emission reduction policies, climate adaptation, and mitigation and sustainable energy policies and initiatives are examined and scrutinized at length. The book provides an overview of select resilience and sustainability strategic planning in communities and organizations. The research highlights organizations and communities that have effectively adopted and implemented resilience and sustainability planning, like the City of Tallahassee in Florida. The private sector perspective is offered for additional insight. Furthermore, the most up-to-date research on COVID-19 pandemic responses by communities and organizations and climate change reports and plans are included in this book.

Chapter 1 provides expanded definitions of strategic resilience and sustainability and the mechanisms reshaping organizations and communities. Chapter 2 examines the strategic planning processes for organizations and communities and determines sustainability and resilience planning steps. Chapter 3 presents the insights into identifying community and organizational level engagement, internal and external stakeholders, organizers, partnerships, collaborators, and implementers of distinct planning stages. Chapter 4 defines the measurements, tracking tactics, observing, screening, improving, and reporting methods using the Quadruple Bottom Line procedure. It presents an example of a resilience and sustainability progress report to ensure accountability, answerability, transparency, and good governance.

Further, Chap. 5 details the implementation of a sustainability and resilience plan once established, describing programs and initiatives already underway in communities and organizations to achieve sustainable and resilient organizations and communities. Chapter 6 examines the theoretical and the practical intersection between resilience and sustainability, and climate change. Chapter 7 evaluates the resources and tools available for resilience and sustainability planning to aid organizations and communities. Chapter 8 addresses the current and future resilience and sustainability in cities, counties, and organizations, including concerns over climate change, pandemics, disaster resilience, and emergency preparedness.

The book introduces the concepts of sustainability, resilience, climate change, and strategic planning methodologies. It contributes to practical solutions and methods in applying resilience and sustainability in strategic planning for organizations and communities. Long-term strategic resilience and sustainability plans are paramount to the longevity of organizations and communities. This book offers the essential planning tools for organizations to develop strategic resilience and sustainability plans. It will benefit public administrators, managers, elected and appointed officials in cities, counties, and townships, educators, managers, and staff in nonprofit and nongovernment organizations. Strategic resilience and sustainability planning advances effectiveness, efficiency in organizations, and resilience in communities in concentrating and institutionalizing strategic planning by addressing good governance, environmental concerns, social problems, and sustainable economic growth.

The target audience is a diverse group of professional staff, officials, administrators, leaders and managers divided into four categories:

Elected and appointed local government officials, city and county planners, administrative analysts, chief administrative officers, chief operation officers, chief financial officers, county administrators, city managers, sustainability directors, chief resiliency officers, budget and financial directors, and community leaders.

University administrators, presidents, provosts, and sustainability directors.

Managers and staff of private and nonprofit and nongovernment organizations.

Professors, academics, teachers, and educators studying and teaching sustainability and resilience applications.

The book aims to answer the following underlying postulates and inquiries.

- The strategic resilience and sustainability plans enhance and support the long-term success of organizations and communities.
- The successful examples of resilience and sustainability initiatives.
- The steps to creating and implementing resilience and sustainability strategic plan.
- Benefits to communities and organizations from having reliable and robust strategic resilience and sustainability plans.
- Measuring, tracking, observing, controlling, and reporting outcomes of the plans.

This book draws on experiences in managing and implementing sustainability and resilience-related activities. It draws heavily from examining climate change, climate preparedness, greenhouse gas emission strategies, climate adaptation, climate mitigation, and sustainable energy policies and plans. It also builds on years of researching the best practices in select cities, counties, townships, and organizations. Best practices and emergent practices from communities and organizations are presented throughout the book as resilient communities' spotlight.

The intersection of sustainability into resilience and logical progression into strategic resilience and sustainability plans are a result of years of municipalities imbuing climate change data, the greenhouse gas emissions mitigation, resilience, and the aspects of adaptation and reduction of the impacts of climate change into strategic planning. Resilience is the ability of organizations to withstand pressures, recover, and continue operating despite disasters, natural or human-made, changing shifts in economic cycles, where climate and extreme weather resilience are expressed as the strength with which cities and regions recover from catastrophe exacerbated by changing climate, global warming, rising seas, and extreme weather patterns. The recent COVID-19 crisis exposed the need for more aggressive resilience planning and policies.

Strategic resilience and sustainability planning improves the quality of service delivery and increases the ability of organizations to respond to disasters and recover effectively. Throughout the book, the terms resilience and sustainability initiatives and strategic planning are explained interchangeably and interdependently. Ideally, communities and organizations adopt a Quadruple Bottom Line in resilience and sustainability planning to transition to a single strategic plan with both approaches utilized. Regardless of the terminology used to describe plans, the ultimate goal of strategic resilience and sustainability planning is to achieve sustainable and resilient organizations and communities.

The book discusses appropriate literature to provide context for resilience and sustainability strategic planning. Administrative professionals may follow the outline and processes to create and then fully implement a strategic resilience and sustainability plan. Such a planning process brings cohesiveness to organizations and agencies to better manage their projects, policies, and programs. Resilience and sustainability planning represents the next stage in strategic planning and is reinforced with climate adaptation, mitigation, and preparedness actions and strategies to achieve resilience and sustainability.

Pensacola Beach, FL, USA
July 11, 2021

Haris Alibašić

"Dr. Alibašić's book is an immense contribution to the study on climate resilience and sustainability, and it must be translated into as many languages as possible. The research, practical, and writing quality and application of knowledge presented in this book need to be replicated everywhere, including in Bosnia and Herzegovina."

—Dr. Senadin Lavić, Professor, Master's in Political Science, Director, *College of Political Science, University of Sarajevo, Sarajevo, Bosnia and Herzegovina*

"The main contribution of the work by Dr. **Alibašić** is the introduction of a Quadruple Bottom Line (QBL). Whereas previous research has focused on the three pillars of economic prosperity, social equity, and environmental integrity (Triple Bottom Line [TBL]) to achieve resilience and sustainability outcomes, Dr. Alibašić masterfully outlines how governance should be added to create a QBL.

The proposed QBL approach examines sustainability issues from the added perspective of focusing on governance. The expanded QBL definition outlines organizational capacity to incorporate and implement definitive policies and programs."

—Dr. Lachezar G. Anguelov, *The MPA Faculty, The Evergreen State College*

"Haris Alibašić has written a masterful primer for those interested in sustainability and resilience. The concise and factual review of the roles local government portrayed in this book is a must-read for any elected and appointed official in counties, cities, townships, and villages. From practical to theoretical, the author weaves an important narrative on why cities and their leaders must engage in sustainability and resilience planning. For those who live and breathe city planning, this book offers a greater understanding of how we implement and measure the quality of life and sustainable improvements in our cities, preparing them for the worst and the best of times ahead."

—Honorable John Dailey, Mayor, *The City of Tallahassee*

"There is so much to like in this book, as it creates an excellent and easy-to-follow framework for implementing sustainability planning in local government. The book expands the traditional definition of sustainability (environmental-economic-social) to include governance—thus expanding it to the "quadruple bottom line." This makes great sense, as our local governments are where communities come together to address their most basic issues, and that can't happen without good governance.

The inclusion of governance in the resilience and sustainability framework, especially local government, is particularly timely, considering that such planning work has recently been deemphasized at the federal level. At the same time, many cities push forward with strong sustainability and resilience planning.

Dr. **Alibašić** is a university professor, but he has made the book accessible to anyone interested in sustainability planning. It's written not as an academic treatise but as a practical guide, taking one step-by-step through the steps necessary to integrate sustainability and resilience planning into a local government. Having read this book, I feel ready to implement sustainability and resilience planning in my community more formally, keeping it handy for easy reference."

—Christian Wagley, *Healthy Gulf Coastal Organizer*

"In *Strategic Resilience and Sustainability Planning*, Dr. Alibašić offers a well-considered view of sustainability in practice for local governments. His addition of governance to the economic/financial, social, and environmental features that typify the triple bottom line is especially welcome. There is a great deal of interest in sustainability, and the role of governance in the prospect of achieving sustainability is often misunderstood. This book will be of interest to local governments engaged in sustainability planning, as well as the college/academic audience."

—Dr. Christopher Atkinson, Assistant Professor, *University of West Florida*

Acknowledgments

I am thankful for the opportunity to have worked with so many dedicated city employees, elected and appointed officials, community leaders, local government administrators, and friends and coworkers in the past. I am grateful to the Florida League of Cities' staff and the colleagues and students at the University of West Florida. I am thankful to Springer's staff for their guidance and encouragement. I am forever grateful to everyone not mentioned but who contributed to the successful completion of this book.

Contents

About the Author

Dr. Haris Alibašić is an Associate Professor and Program Coordinator in the Public Administration program at the University of West Florida. Dr. Alibašić brings 27 years of expertise and experience in the public sector, including working for the United Nations Mission and the Office of High Representative in Bosnia and Herzegovina and directing energy, sustainability, and legislative affairs policies and programs for Grand Rapids, the second-largest city in Michigan. In Grand Rapids, he promoted sustainable policies. He implemented resilience strategies resulting in significantly reduced energy usage and cost and spurring significant renewable energy investments and development of climate resilience plan for the region.

For over 17 years, Dr. Alibašić taught graduate and undergraduate courses in public policy, public administration, economic development, emergency preparedness, management, and sustainability at Grand Valley State University, Central Michigan University, and Davenport University. As an associate professor at UWF, he teaches doctoral and graduate-level online and in-class courses in public service and administrative ethics, the political economy of public administration, strategic management in administration, leadership in sustainability, public budgeting and finance, capstone, and public administration.

Dr. Alibašić is a partner in the Florida League of Cities Municipal Research Program. Through the Partners in Municipal Research Program, the Center for Municipal Research & Innovation serves as a link between Florida's public policy researchers and municipal governments, bridging the gap between academics and public policymakers and administrators. For the past 3 years, he led the UWF interdisciplinary team of undergraduate and graduate students to the statewide Municipal Modernization competition organized by the Florida League of Cities and held in Orlando, FL. He presented at the FLC symposium on climate resilience and hosted a webinar on climate resilience and sustainability planning. Dr. Alibašić is an elected member of the Executive Committee of the American Society for Public Administration (ASPA)—Section on Ethics and Integrity in Governance (SEIGOV), where he serves as the Director of the Charter, ByLaws, and Legal Afffairs.

In March 2017, Dr. Alibašić was appointed by the City Council to the City of Pensacola Climate Mitigation and Adaptation Task Force at the recommendation of local community members and an elected official. On November 8, 2018, Dr. Alibašić presented the Task Force's final recommendations to the City of Pensacola Council. In December 2018, *he was interviewed for a*

WUWF radio segment on climate change in the region and discussed the Task Force climate resilience report. In September 2020 and then again in June 2021, Dr. Alibašić was invited to speak on the topic of the climate resilience report and recommendations with the City of Pensacola administrative leadership and the Environmental Advisory Board for the city.

In 2013, Dr. Alibašić advised the Resilient Communities of America on climate resilience and went on to serve as co-chair for the energy sector of the White House Climate Preparedness and Resilience Task Force in 2014, helping to research and write several sections of the task force report and recommendations to President Obama.

Dr. Alibašić received the 2012 West Michigan Environmental Action Council (WMEAC)—The C.R. Evenson Award and the 2011 Grand Valley State University's Sustainability Champion Award. In November 2016, he won the prestigious Sustainable Hall of Fame Merit Award from West Michigan Sustainable Business Forum. In January 2017, he received an Emerging Scholar Award at the Thirteenth International Conference on Environmental, Cultural, Economic and Social Sustainability and the On Sustainability Research Network, held in Rio, Brazil. He also received a 2017 Great Lakes-Saint Lawrence Cities Initiative (GLSLCI) Certificate of Appreciation and November 28th of 2016, the State of Michigan Special Tribute. In 2018, Dr. Alibašić received the CEPS Outstanding Contribution in Research Award.

Dr. Alibašić has written, presented, and published extensively on the topics of ethics, sustainability, climate resilience, economic development, and sustainable energy. He is the author of the book *Sustainability and Resilience Planning for Local Governments: The Quadruple Bottom Line Strategy*, published in June 2018 by Springer, with the second edition of the book expected to be published in 2022.

Dr. Alibašić presented at numerous national, regional, and international conferences on sustainability, resilience, economic development, and ethics, including in Brazil, Costa Rica, Monte Negro, Italy, Austria, Croatia, and Bosnia and Herzegovina. In 2021 alone, Dr. Alibašić presented at three international virtual conferences held by the American Society for Public Administration, International Small Islands Studies Association, and at the Seventeenth International Conference on Environmental, Cultural, Economic and Social Sustainability hosted by Vrije Universiteit Amsterdam in the Netherlands. He has been interviewed and quoted in local, national, and international media for his expertise and research in sustainability, resilience, and ethics.

Research Interest

- Climate Resilience and Sustainability Planning
- Ethics, Integrity, Administrative Evil, Moral Inversion
- Sustainable Energy, Renewable Energy, and Energy Efficiency
- Corporate Social Responsibility
- Resilient Emergency Preparedness and Disaster Response

Dr. Alibašić is an International Editorial Board member for *Pregled*, a peer-reviewed journal of social science studies at the University of Sarajevo, Bosnia and Herzegovina. He served as a Section Editor for the *Creighton Journal of Interdisciplinary Leadership* and a regular peer-reviewer for the *International Journal of Climate Change: Impacts and Responses and for the Energies, Sustainability, Journal of Energy Policy.* From 2018 to 2019, he served on the Executive Review Board and as a Digital Media Strategist for *Public Integrity*, a double-blind peer-review journal on ethics and integrity. Together with Dr. Jonathan Rose, Dr. Alibašić is a co-editor for the special issue of the Public Integrity symposium "Fake News, Post-truth, and Alternative facts: Lying and Integrity in the Public Sector."

Degrees and Institutions: Dr. Alibašić holds a Bachelor's degree in Business Administration (BBA) in International Business and Marketing and a Master's degree in Public Administration (MPA) from Grand Valley State University (GVSU) in Grand Rapids, MI. He earned a Ph.D. in Public Policy and Administration from Walden University in Minneapolis, MN, where he was a recipient of the Doctoral Scholarship, Commitment to Social Change.

Dr. Alibašić is actively involved with the Bosnian American community and served as a past president of the Congress of North American Bosniaks (CNAB). He is a past board member of the Advisory Council for Bosnia and Herzegovina, an international expert team member of the Institute for Research of Genocide (Canada). He received a North American Bosniaks' Special Recognition Award in 2013 for outstanding contributions to the advancement of Bosniaks and Bosnia and Herzegovina.

Initiating and Assessing Strategic Resilience and Sustainability Planning

1

"At the hear of a learning organization is a shift of mind-from seeing ourselves as separate from the world to connected to the world, from seeing problems as causes by someone or something out there to seeing how our own actions create the problems we experience. A learning organization is a place where people are continually how they create their reality. And how they can change it."

Senge (2010, p.12).

Overvivew

The first chapter answers inquiries about strategic resilience and sustainability planning and how it reshapes communities and organizations. In this chapter, exact terms and definitions related to strategic resilience and sustainability planning, the history of sustainability, climate resiliency, climate change, Triple Bottom Line (TBL), Quadruple Bottom Line (QBL), greenhouse gas emissions, carbon footprint, and climate resilience and preparedness are presented. The chapter develops definitions and terms related to resilience, sustainability, and the tools reshaping communities and organizations. Also, a careful review of successful strategic resilience and sustainability plans is offered. An analytical assessment of the Quadruple Bottom Line is proposed, focusing on governance as the vital element. The book brings perspectives on private sector resilience and sustainability planning and lessons from COVID-19 responses.

Keywords

Sustainability · Resilience · Climate resilience · Climate preparedness · Climate change · Triple Bottom Line (TBL) · Quadruple Bottom Line (QBL) · Strategic planning · Sustainability planning · Resilience planning · Sustainable energy · Local governments · Cities · Communities · Climate adaptation · Climate mitigation · Global warming · Carbon footprint · Greenhouse gas emissions

Key Questions

This first chapter of the book is aimed at answering the following underlying assumptions and inquiries:

- What is sustainability?
- What is resilience?
- What is resilience and sustainability planning?
- How do resilience and sustainability enhance and support the long-term success of organizations and communities?
- How do resilience and sustainability plans work with the organizational values, missions, and goals?
- What is the resilience and planning process?
- How do cities, municipalities, communities, and organizations benefit from resilience and sustainability initiatives?

H. Alibašić, *Strategic Resilience and Sustainability Planning*, Sustainable Development Goals Series, https://doi.org/10.1007/978-3-030-91159-1_1

This chapter provides expanded and advanced definitions of sustainability and resilience and the tools for reforming local government organizations and communities. Resilience and sustainability increase effectiveness, efficiency in organizations and institutionalize robust strategic planning by addressing good governance, environmental concerns, social problems, and sustainable economic growth.

Introduction

In this chapter, particular definitions linked to resilience and sustainability are explained. The history of sustainability and Triple Bottom Line (TBL) and its intended transformation to Quadruple Bottom Line is explained. Additionally, the terminologies on greenhouse gas emissions, carbon footprint, climate resilience, climate disaster, resilient emergency preparedness, and resilience and sustainability planning are reviewed. The topic of sustainability and resilience and their effects on local governments, communities, and communities justifies continuous investigation, study, and in-depth analysis. The COVID-19 pandemic brought to light the relevance of resilience and sustainability planning. The COVID-19 world faces a more intense need for planning for resilience and sustainability to face the next conceivably imminent world crisis.

An invigorated definition of sustainability and resilience, including an extended review of the contemporary and relevant literature on sustainability, resilience, and climate change, is included in the chapter. Also, a judicious review of prosperous resilience and sustainability strategic plans in select local governments is added. The central aspect of resilience and sustainability planning is its connection to the Quadruple Bottom Line elements. Communities and organizations have faced increased pressures, challenges, and scrutiny during the COVID-19 pandemic. Local governments' responses vary on their level of preparedness and commitments to resilience and sustainability.

The book exhibits the evolution from the Triple Bottom Line to the Quadruple Bottom Line, emphasizing governance. The topics of resilience and sustainability are under unremitting reflection, scrutiny, and inspection. Resilience and sustainability are often described to guarantee the positive impact of actions undertaken by organizations or individuals. Infrequently, sustainability and resilience are mistaken for only the financial impact on organizations and their projects and operations.

The chief drivers of robust resilience and sustainability plans are the capacity of communities and organizations to adapt to the environmental, societal, and economic circumstances surrounding them. Local governments, communities, and organizations use sustainability and resilience to address obligations and responsibilities to their constituents' demands, engaging in innovation to deliver quality of life services as funding and revenues shrink. Local elected and appointed leaders are aware of the complex nature of cities and design programs to address cities' sustainability needs and enhance the resiliency of cities in response to threats, including emergencies, disasters, extreme weather, and climate change. An effective sustainability and resilience plan can help tackle those apprehensions, including pandemic readiness. Resilience is a useful construct to demonstrate the likelihood of recovery, as Matarrita-Cascante et al. (2017) remarked and noted by Cafer et al. (2019) on the positive community changes. Magis (2010) observed resilience in social sustainability as the capacity of systems to adapt and transform.

Definitions and Terms

Describing resilience and sustainability and using appropriate terminology are quintessential to accepting the concepts and further improvements of the strategic planning process. The following are some of the most relevant terms used throughout the book.

Carbon Footprint The total volume of greenhouse gas emissions (GHG) produced by an organization (or an individual) in a given period due to the production and consumption of traditional sources of energy.

Climate Change A global change and shift in the average temperature and seasons, coupled with the increased unpredictability and extreme nature of otherwise established weather patterns. It is a scientifically supported fact proven by numerous studies and research showing the occurrence on a global scale, linked to human actions of the exploitation of traditional sources of energy.

Climate Adaptation Community, regional or national strategies deployed by public and private sector organizations to adapt to the changing conditions resulting from climate change.

Climate Mitigation Strategies and activities exercised by organizations to decrease the pollutants from the use of traditional sources of energy such as coal, oil, and natural gas.

Climate Preparedness and Readiness An overall strategy deployed by local governments and other organizations to prepare and implement strategies to combat perils stemming from the changing climate and extreme weather.

Climate Resilience An advanced strategy to prepare organizations and communities for threats from climate change and planning for such risks, incorporating various economic, environmental, social, governance, and emergency preparedness strategies into resilience and sustainability planning. Such planning strategies prepare organizations and communities to withstand distresses, shocks and disasters and to continue to function during and after the adversities.

Energy Efficiency Investments and activities made to reduce energy consumption and positively impact organizations and individuals by reducing costs of energy in buildings and operations.

Feed-in Tariff A policy explicitly developed to encourage the use and production of renewable energy by paying an above the cost price to energy producers for renewable energy production.

Global Warming The trend of persistent increases in the Earth's overall temperatures since the 19th century, due to more significant heat-trapping carbon dioxide and other gas emissions resulting from continuous energy production and consumption from coal, oil, and natural gas.

Greenhouse Gas Emissions (GHG Emissions) Emissions of gases trapping heat in the atmosphere, considered greenhouse gases (GHG) which include carbon dioxide (CO_2), methane (CH_4), nitrous oxide (N_2O), and fluorinated gases.

Local Government Resilience The ability of municipalities (cities, counties, townships, and villages) and counties to prepare and react to disasters, respond robustly to adverse situations, and continue operations after the disasters.

Peak Oil A scientific argument made by oil and energy experts predicting that the world either reached its daily oil production capacity or will have reached it soon, predicting a reversed trend in oil production and causing a future energy crisis.

Process improvements A strategy employed by organizations focusing on eliminating waste in the processes and improving efficiency and efficacy of service delivery based on manufacturers' approach to streamlining operations.

Quadruple bottom line (QBL) Organizational capacity to embed and incorporate a set of definitive policies and programs to address sustainability's economic, social, environmental, and governance elements, whereas governance is expressed through fiscal responsibility, community engagement for effective service delivery, transparency, and accountability for more resilient organizations and communities.

Renewable Energy Energy produced from sources that appear infinite in its current form, wind, hydro, solar, and geothermal (power from underground Earth's heat).

Renewable Portfolio Standard (RPS) A policy program usually adopted by states to require energy generators/utilities to produce a certain percentage of electricity from renewable energy sources.

Resilience The ability of organizations to withstand threats, shocks, and negative impacts and recover and continue operation afterward.

Resilience Planning An ability of organizations to plan long-term using system-wide approaches to recover from adverse effects of human and nature-made disasters, including climate change and pandemics, such as COVID-19.

Sustainability An ability of organizations to produce products and deliver services comprehensively with a minimum negative social, economic, environmental, and governance effects on resources while maximizing positive outcomes.

Sustainability Plan An active planning process with governance, economic, environmental, and social outcomes and targets for organizations and communities.

Sustainable energy An ability of organizations and society to efficiently impact their bottom line and provide positive social change through reduction in energy consumption, production of renewable energy, and efficient energy management.

Triple bottom line (TBL): A concept that describes how socially responsible organizations consider the negative impact of their actions on social and environmental aspects of society and try to minimize it by using sustainability in its core mission, values, and operations. The same principle applies across nonprofit and governmental sectors.

Vulnerabilities Weaknesses in the system within quadruple bottom line strategic resilience and sustainability planning framework. Vulnerability assessment identifies weakness in organizations and communities.

Waste Minimization An ability of organizations and communities to reduce the waste designated for landfills by deploying the recycling, reusing, and repurposing processes to minimize waste. Life cycle assessment provides a cradle-to-grave or cradle-to-cradle assessment of the environmental consequences of a given product, including the raw material used in its production, distribution, supply chain, end of use, and reuse.

Climate Change as the *Impetus* for Resilience Planning

The well-documented climate change causes the most serious threats to communities and societies (Fletcher 2013; NASA 2021; UNFCCC 2015a, b). Extreme weather and other associated climate change challenges increase annually. Watts (2021a, b) noted scientists' concerns over the unprecedented scale of floods in Europe and the United States caused by human activities. The complex nature of intergovernmental coordination and action on climate change, the disconnect between the legislative and executive branches of government, lead to less than desired outcomes on the issue of climate change (Center for Climate and Energy Solutions 2021; Federal Resilience Program 2021). There are serious consequences of inaction on climate change and lack of climate preparedness. Local government officials bear responsibility and the brunt of impacts from climate change (Alibašić 2020).

Organizations committed to resilience and sustainability strategic planning protect environmental and social resources by delivering the most efficient services and improving governance. There are connections between the practical and applied aspects of sustainable and resilient actions and organizations' impact on society. By actively pursuing climate resilience investments, organizations positively impact their economic, social, governance, and environmental bottom line, thus affecting positive social change. Sustainability and resilience outcomes are feasible and attainable in cities at local levels of government in their pursuit of policies and programs regardless of political affiliation. At the very least, it may be less controversial to gain consensus at the local level than at the state and federal levels. Sustainability and resilience-related efforts aim to impact an organization's economic, environmental, and social governance.

The Significance of Resilience and Sustainability

Resilience and sustainability are the focal points in explaining the effectiveness of service delivery, disaster responses, emergency management, and climate change preparedness. Sustainability and resilience may be described as the connected social, economic, environmental, and governance concerns within an organizational framework. Sustainability is often mistakenly viewed through financial effects on institutions and operations and is commonly misunderstood as costly. Contextually, organizations aim to apply sustainability in practice to further their operational efficiency and address their actions' governance, economic, environmental, and societal impact. There are parallel impacts of organizational activities on the governance, environment, society, and economy. Also, organizational resilience to the external dynamics and impacts, including the governance changes, economic interruptions, environmental disruptions, and social stressors, is internal and external.

Amidst trepidations over intensified pressures on the cost of energy, climate change, pandemic, and financial conditions on the organization's bottom line in the short and long term, sustainability and resilience planning affords municipalities a chance for advancement. The term "sustainability" conveys a certain sense of continuity that withstands the test of time. Slavin (2011) alluded to this sense of endurance in defining sustainability. Becker (2011) recognized sustainability through morality and system design. The resilience and sustainability strategic planning equally treats the governance, economic, environmental, and social issues. In the pandemic world, it is imperative to treat all four elements with the same intensity.

Additionally, resilience may be described as an enhanced level of planning by considering climate change and other emerging ambiguities. Equally important for modern organizations is expanding the concept of resilience and sustainability and framing it through climate resilience and climate change-related impacts. Resilience adds an organizational and communities level of protection as they face heightened uncertainties, reflected through climate change, pandemics, extreme weather events, infrastructure failures, and human resource pressures.

Sustainable and resilient organizations embrace a common commitment and endorse principals through stakeholders' engagement by focusing on the broad concept and vision of improving social, environmental, economic, and governance benefits. Furthermore, identifying shared goals and objectives will further assist organizations in achieving sustainability and resilience outcomes.

Theoretical Background

Theories associated with sustainability and resilience are varied and diverse and offer a starting point to examine effective strategic planning. There are varying concepts, theories, standards, and approaches used to form a sustainability and resilience framework and corresponding governance, economic, environmental, and social bottom lines for organizations and society. A body of work from other disciplines from economic, social, environmental, leadership, and political science fields provide a meaningful theoretical outline of resilience and sustainability correspondingly.

Heal (1998) explained sustainability as a metaphor for the consequential issues facing humanity. Additionally, Gaertner (2009) viewed the theory of social choice through the lens of a collective approach to the decision-making process in preference of choices in society. Agyeman (2005) viewed sustainable communities in the context of environmental justice. Daly (1996) argued for sustainable economic growth. Analyzing sustainability and related policies offers a better grasp of the measures undertaken and overall outcomes on the society or organizations. Elster and Hylland (1989) argued that social choice theory emanated from the aggregation of collective welfare. The effective approach to measuring sustainability and resilience is to define problems and processes of social welfare through the economic, social, environmental, and governance lens.

Heal (1998) methodically explained the essential premises of environmental assets and the benefits from such assets. The method emphasized the environmental benefit as a basis of sustainability without delving into the social or governance bottom lines. Solow (1992) offered a rational policy approach using economic theory to defend the notion of possible improvements to the economy and the environment. While the focus is on the environment, a clear understanding is that an improved environment leads to enhanced economic and societal outcomes.

In addition to social choice theory and economic theories, another theoretical framework connected to sustainability, is the system theory. Introduced by von Bertalanffy (1969/2013), system theory classifies all actions as interconnected within a bound system. Von Bertalanffy (1950) deliberated on how general system laws apply to all elements within the confines of a system, and the general system theory applies to all sciences dealing with systems. Any phenomenon may be regarded as an interconnected system of different elements, from sustainable energy processes to climate change readiness. Patton (2002) maintained the entirety of the system is greater than each part of it. Such an approach facilitates explaining the conflicted sustainability and resilience phenomena and the methods underlying elements of environmental, social, governance, and economic components function within such a framework. Recently, Fernández-Prados et al. (2021) and Bryce et al. (2020) used the concept of resilience to explain the ability of the populace to contend with the unprecedented risk and uncertainty associated with COVID-19 pandemic.

Scientists and researchers supported including resilience in emergency preparedness, disaster responses, and crisis plans in recent years. Son et al. (2017) and Tveiten et al. (2012) adequately discussed the incorporation of resilience in emergency administration and disaster response. Moreover, Koch et al. (2016) contended that the community risk and vulnerability assessment strengthened with resilience.

Sustainable Development and Triple Bottom Line (TBL)

In 1969, the United States Congress passed the National Environmental Policy Act with the following language "recognizing the profound impact of [hu]man activity on the interrelations of all components of the natural environment, particularly the profound influences of population growth, high-density urbanization, industrial expansion, resource exploitation, and new and expanding technological advances and recognizing further the critical importance of restoring and maintaining environmental quality to the overall welfare and development of [hu]man, declares that it is the continuing policy of the Federal Government, in cooperation with State and local governments, and other concerned public and private organizations, to use all practicable means and measures, including financial and technical assistance, in a manner calculated to, foster and promote the general welfare, to create and maintain conditions under which man and nature can exist in productive harmony and fulfill the social, economic, and other requirements of present and future generations…." (The Congress 1969). The intent is to recognize and react to the environmental and societal harm from unprecedented human activities associated with exploiting and destroying natural resources. This could be viewed as one of the early steps of resilience and sustainability planning. The policy calls for the continuing responsibility of the Federal Government to use all practicable means, consistent with other essential considerations of national policy, to improve and coordinate Federal plans, functions, programs, and resources to the end that the Nation may, among other things, "achieve a balance between population and resource use which will permit high standards of living and a wide sharing of life's amenities, and enhance the quality of renewable resources and approach the maximum attainable recycling of depletable resources" (The Congress 1969).

Another foundation of sustainability and sustainable development were ascertained by the United Nations World Commission on Environment and Development. The United Nations World Commission on Environment and Development (1987) coined the term sustainable development as the responsible administration of resources in the present by organizations and individuals without compromising the needs of future generations. The Report of the World Commission on Environment and Development: Our Common Future from Brundtland Commission was set up by the United Nations, establishing the unique classification of what sustainability means as the development that looks to meet the needs of today's population without jeopardizing the future generations' economic and social opportunities.

The United Nations World Commission on Environment and Development (1987) established the sustainable development framework considering future societal needs. However, since the initial platform for sustainability was developed, much research was advanced in redefining and refining sustainability. There are inconsistent interpretations of sustainability and its impact on organizations, communities, and society. Additionally, Stubbs and Cocklin (2008) noted the lack of consensus on the definition of sustainability.

Elkington's (1997) pioneering and groundbreaking views on sustainability tasked the corporations to evaluate their environmental and social impact. Elkington's (1997) classic offered the Triple Bottom Line approach to positively influence companies and organizations. The economic, social, and environmental elements considered under the proposed framework are often referred to as the Triple Bottom Line (TBL) (Elkington 1997; Savitz and Weber 2006).

From Triple Bottom Line (TBL) to Quadruple Bottom Line (QBL)

The early concept of sustainability was defined through the Triple Bottom Line approach to measuring impact from organizations on society as the three elements of economic prosperity, social equity, and environmental justice as a balanced value proposition. The Triple Bottom Line axiom asked the corporations to evaluate their social and environmental impact on society and the environment beyond what they produce for their economic benefit. Sustainability and resilience represent an opportunity for the advancement of organizations and the necessary evolution of society. In his later writings, Elkington (2012a, b) posited that sustainability supports better corporate governance to sustain capitalism. The Triple Bottom Line definition focuses on the private sector, and it broadly applies to public sector organizations. The imperative is to invite the private sector organizations to implement goals focusing on economic prosperity, environmental protection, and social equity. However, as sustainability evolves, its static description looking through three basic elements of sustainability needs constant reinvention and revisiting.

The advanced Quadruple Bottom Line examines the issue of sustainability and resilience from an added perspective of governance. Alibašić (2017, 2018f) offered an expanded definition of Quadruple Bottom Line through the capacity of organizations "to embed and incorporate a set of definitive policies and programs to address economic, social, environmental, and governance aspects of sustainability, whereas governance is defined through fiscal responsibility and resilience, community engagement for effective service delivery, and

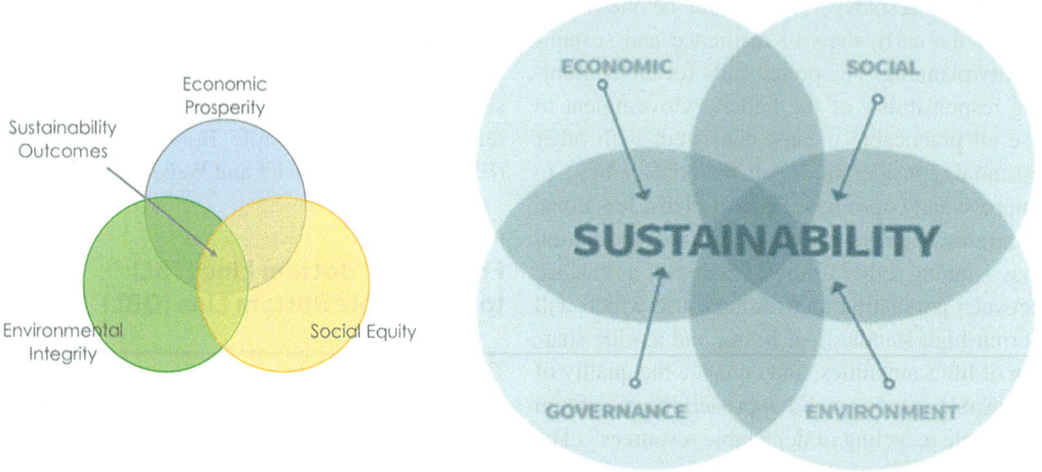

Fig. 1.1 An evolution from Triple Bottom Line to Quadruple Bottom Line

transparency and accountability" and more resilient and sustainable organizations and communities.

Governance is a dynamic component critical to the success of resilience and sustainability. In analyzing transition management, Loorbach (2007) realized the relevance of governance for sustainability. Including and assessing good governance is decisive to the evolution of sustainability and resilience. Gray and Milne (2002) saw a dual role for the public organizations as either contributing to improvements or worsening social, economic, and environmental conditions. In the words of Bryson (2004), public sector organizations create public value. Addressing governance aspects of public sector organizations increases the likelihood of creating more efficient value creation.

Visseren-Hamakers (2018a, b) discussed integrative governance in the context of environmental policies and the roles actors and institutions have in governing. Ostrom (1990) analyzed the collective action in governing the commons. Organizations with the core value of good governance are transparent and accountable, relying on answerability to measure progress. Figure 1.1 diagram explains visually the transition from Triple Bottom Line to Quadruple Bottom Line.

Resilience and Sustainability of Local Governments

Early roots of the local government and sustainability are traced to the United Nations Conference on Environment and Development in 1992 and Report of the United Nations Conference on Environment and Development Vol. 1, with Resolutions adopted by the Conference (Agenda 21) in Rio de Janeiro, Brazil. The Agenda 21 report acknowledged the major environmental issues facing fast-growing cities and the number of cities, calling "for greater attention to issues of local government and municipal management" (United Nations Conference on Environment & Development 1992, p. 1).

Among the key aspirations of Agenda 21 was "to implement policies and strategies that promote adequate levels of funding and focus on integrated human development policies, including income generation, increased local control of resources, local institution-strengthening, and capacity-building, and greater involvement of non-governmental organizations and local levels of government as delivery mechanisms" (United Nations Conference on Environment & Development 1992, p. 14). Agenda 21 brought to light social equity and nature in economic development. The well-being of humans and harmony with the environment are at the center of Agenda 21, Principle 1. Balancing social equity with a clean, healthy environment, and a sound economy creates sustainability and warrants resilience. Conversely, ignoring the needs of the most vulnerable and impoverished populations creates devastating effects on society.

As cities shift their attention to addressing social equity in their sustainability and resilience plans, the definition of social equity from the public sector perspective means that all its residents have equitable access to services, education, training, learning, health care, and resources for a high-quality life. Public service organizations distribute services fairly and equitably, promoting justice and equity in developing and implementing public policy and programs. The goal of resilience and sustainability strategic planning efforts is to impact the governance, economic, environmental, and social aspects of an organization.

With the increasingly global scope of today's world, collaboration is the key to success. More tangibly, elements that create sustainable communities include local government support and cohesion on land zoning ordinances, water systems, transportation systems, waste management systems, natural resources preservation, and food production (Coyle 2011). The UNFCCC's (2015a, b, p. 9). Paris Climate Agreement recognizes the importance of cities and counties and includes contributions to resilience and sustainability in the final document, including Article 7:

> Parties recognize that adaptation is a global challenge faced by all with local, subnational, national, regional, and international dimensions, and that it is a key component of and makes a contribution to the long term global response to climate change to protect people, livelihoods and ecosystems, taking into account the urgent and immediate needs of those developing country Parties that are particularly vulnerable to the adverse effects of climate change.

Resources imperative to the sustainability and resilience of communities are energy, wind power, solar power, transportation, health, education, and public safety. Intangible resources include collaboration between government, population, the private sector, and the nonprofit sector, and that cooperation and excitement will drive the process. Saha (2009) appraised the rise of prominence of sustainability in recognition of the effectiveness of local actions. Local governments found the need to fill the void left by national governments in meeting their sustainability and resilience objectives.

In such an environment, communities, and organizations face challenges due to the world's dependence on traditional energy. Coupled with the fact that contemporary cities' functions are different from those of the past, different threats and opportunities occur daily. Girardet (1999/2006) offered a further understanding of a sustainable city, identifying its enormous impact on the economy and environment through positive actions to reduce energy demand and energy consumption.

Local government organizations provide services and meet increased demand for services while facing constant and severe budget cuts to staffing and operations. At the same time, city and county governments are expected to provide the same level of services without additional revenues or resources. Institutionalizing resilience and sustainability is an enormous undertaking that requires leadership and readiness to measure, track, observe, and report progress and results. With available funding in peril, the local government's capacity to pursue resilience and sustainability initiatives is also at risk. For resilience and sustainability to be fully embedded within organizations, it must become an integral part of the budgeting process by actively pursuing sustainability goals and targets.

Cities and counties contend with constant and severe budget cuts to staffing and operations. Similarly to Martin et al. (2012), authors Ammons et al. (2012) discussed the new normal for local governments' service provisions. Authors argued the long-lasting impact on local governments and their delivery of services beyond the recent economic recession and downturn in the economy (Martin et al. 2012; Ammons et al. 2012). At the same time, city and county governments are expected to provide the same level of services without additional revenues or resources. Institutionalizing resilience and sustainability within organizations is an enormous undertak-

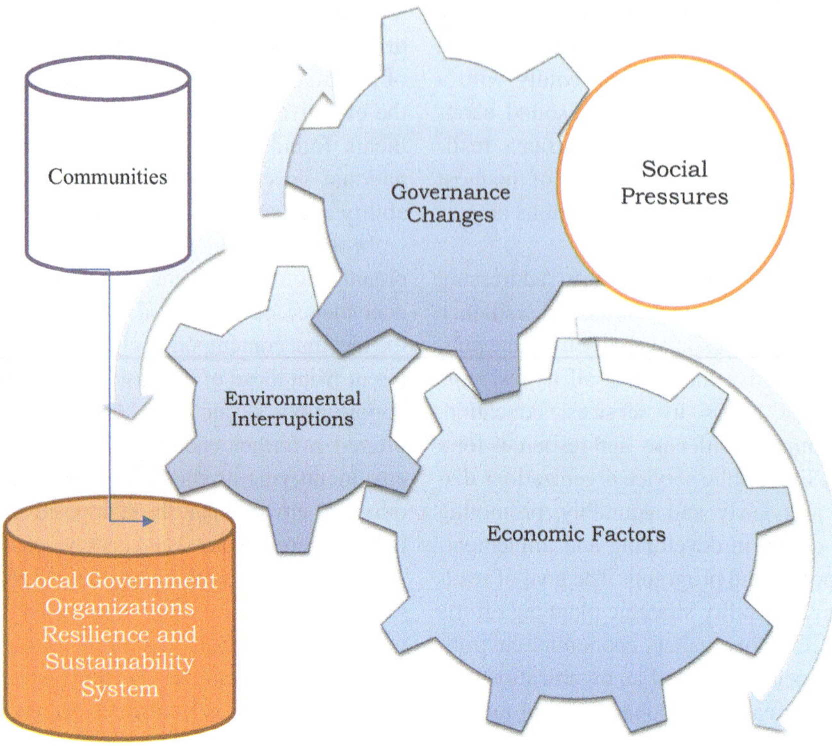

Fig. 1.2 Local government organizations and communities in the resilience and sustainability system boundary

ing, which necessitates leadership, and readiness to take corrective measures. While lamenting federal inaction on climate change, Fitzgerald (2010) praised local government efforts in linking sustainability to climate change threats. Local governments' capacity to pursue resilience and sustainability planning is an integral component of the system boundaries. Figure 1.2 describes the system boundary of local government organizations, communities, and elements manifested in the resilient and sustainable systems.

Private Sector Perspective: Resilience and Sustainability

The pursuit of resilience and sustainability is not constrained to the public sector alone. In making the business case for sustainability for businesses, Werbach (2009) noted the rapid changes in the global economy. Furthermore, Waddock (2009) discussed corporate sustainable practices, including product accountability, life cycle management, spreading the cost of emissions, and benchmarking them with others. Alibašić (2018a, b, c) examined the design of resilient organizations and the role of local corporations in addressing climate change. Strategic resilience and sustainability are observed through a lens of long-term implementation strategy and initiatives by a given organization, with the ultimate objective of providing sustainable and resilient services and products. Davis et al. (2018) examined the connection between corporate performance, green ranking, and sustainability.

Sustainable organizations strive to deliver the most positive economic and societal impact with the least negative impact on the environment with effective governance. Companies realize and find the case for sustainability in knowing the risk of failed cities, communities, and ultimately social infrastructure on their operations. Corporations adjust and seek opportunities for long-term solutions while deploying sustainability and resilience. On a large scale, problems facing companies and urban centers appear to be interconnected to the opportunities and woes of the global economy.

As the societal paradigms shift, corporations, local governments, and organizations utilize sustainability and resilience to support their long-term strategic goals. Hardjono, Van Marewijk and de Klein created The European Corporate Sustainability Framework (ECSF) and (as cited by Stubbs and Cocklin 2008) developed tools for environmental sustainability. Hardjono et al. (2004) provided a comprehensive methodology for implementing sustainability in organizations by capitalizing on organizational dynamics. In utilizing a systematic approach, companies employ corporate sustainability and social responsibility to advance the company's interests further. Savitz and Weber (2006) defined the Triple Bottom Line as the factor of sustainable businesses, arguing sustainable companies anticipate risks and sustainably plan for each.

Sustainability challenges corporations to assess their social and environmental impact on society, beyond the economic effect, as a collective good, and a maximized business opportunity (Galea 2004, p. 37; Elkington 1997). Corporations use sustainability and resilience planning to break ranks with the accepted views of businesses interested in profit only. Such companies advocate for societal issues, including climate change and the protection of the environment.

Bendell and Kearnis (2005) argued that corporations are integral to the resilience and sustainability framework. Furthermore, Benn et al. (2014) discussed how companies use their economic and business clout to pursue political agendas and advocate for resilience and sustainability-related policy changes, impacting markets and society. Baumgartner and Rauter (2017) focused on corporate sustainable development. There is a growing effort by corporations to address climate change and the expectation from the public they will continue engaging in climate resilience activities (Mufson 2019). Corporations not addressing climate change are often rebuked (Murray 2019; Pacthod and Pinner 2021). Conclusively, sustainability and resilience demand a collective, collaborative effort by a broad segment of the population, including a commitment by public, nonprofit, and private sectors.

As depicted in Fig. 1.3, the private sector's successful integration of strategic resilience and sustainability elements depend on multiple fac-

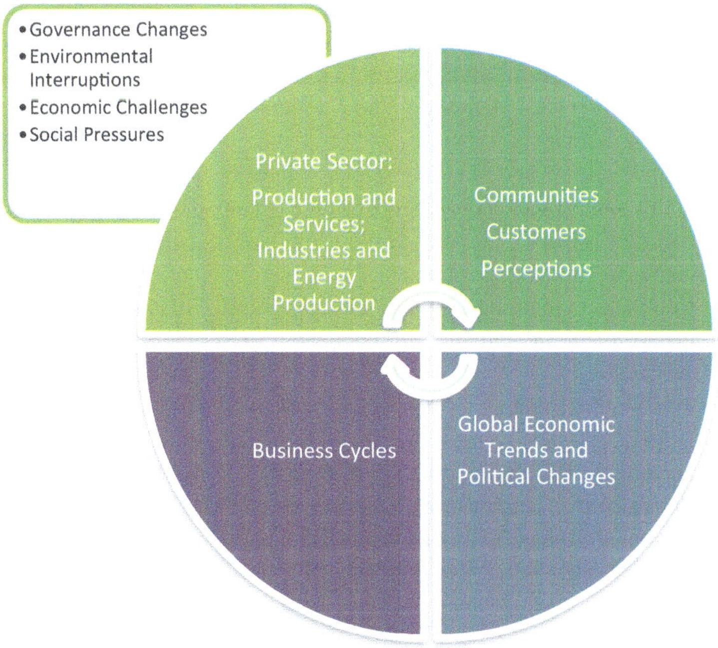

Fig. 1.3 Private sector's resilience and sustainability planning system boundaries

tors and are defined through changes in governance, environmental interruptions, economic challenges, and social pressures. In its system boundaries, businesses contend with a myriad of issues, including regular business cycles, global economic unpredictability, customers perceptions, and changing trends, encompassed in the resilience and sustainability factors.

Commencing Resilience and Sustainability Planning

Similar to strategic planning, creating a resilience and sustainability plan is not an apparent and straightforward task. Developing a strategic resilience and sustainability plan is an organic, bottom-up, linear, and engaging process. Successful organizations integrate resilience and sustainability into strategic and budgetary planning. Sustainability and resilience planning include the practical approaches concerning value added to the organization and the community. The process of embedding sustainability and resilience at all levels of organization involves long-term commitment with a complete engagement. Several elements contribute to the successful application of sustainability and resilience efforts, including but not limited to:

- Improved operational internal efficiency using service delivery and efficiency improvements;
- Notable policy and planning stages—sustainability plan, renewable energy goals, green building policy, climate resilience plans, emergency management plans;
- Complete community's participation and engagement in pursuit of resilience and sustainability;
- Positive engagement of staff and key stakeholders and involvement of all elected and appointed officials are critically relevant;
- Purposefully empowering employees to champion sustainability and resilience targets and outcomes;
- Measuring, tracking, adjusting, observing, and reporting results of the resilience and sustainability-related efforts, using progress reports. As part of such efforts, setting clear goals and objectives is imperative, coupled with specific targets of the plan;
- Carefully connecting targets to the budget and financial plans and policies. Each segment of creating a sustainable and resilient community is linked.

At a minimum, the starting position for strategic resilience and sustainability plan should include the following list of questions.

- What are the organizational and community-wide governance, economic, environmental, and social issues?
- What projects and issues have the potential for the most impact from that organization's perspective?
- How do sustainability and resilience align with organizational; goals, vision, and mission statements?
- Does the organization measure its greenhouse gas emissions and carbon footprint? If so, when was the report last updated?
- Who are the principal stakeholders in the planning of sustainability and resilience strategies? Who are the internal or external stakeholders, partners, and collaborators?
- What are the relevant steps to engage stakeholders?

Climate Resilience Planning

The recent pandemic of COVID-19 revealed the full spectrum of vulnerabilities of the local population in cities. In addition, climate change represents one of the most critical issues that cities are facing. While almost all scientists agree on the causes and impacts of climate change, the inaction on the federal level has left many municipalities to deal with this existential threat on their own. The perils differ across regions and countries, and the ongoing studies of climate change indicate explicit threats to cities around the world and in the United States (IPCC 2014; USGCRP 2014; UNFCCC 2015a, b; NASA 2021).

In 1997, the Kyoto Protocol was adopted as an international agreement to commit participating countries to reduce greenhouse gas emissions (UNFCCC 2014, 2015a, b). The United States did not endorse the Protocol. The previous administration withdrew from the Paris Climate Accords (Alibašić 2018d, e).

However, with the election of President Biden, the US administration expressed its intent to return to Paris Climate Accords (The White House 2021a). Building on President Obama's climate preparedness work (Task Force 2014), President Biden took the systematic approach to address the climate crisis, establishing the White House Office of Domestic Climate Policy to coordinate and implement the President's domestic climate agenda and establishing the National Climate Task Force, with 21 federal agencies and departments working on climate agenda (The White House 2021b). The US administration appointed former presidential candidate John Kerry as a Presidential Envoy for Climate, a member of the United States National Security Council (NSC), to lead the US climate efforts (The White House 2021b, c). For many communities, the US leadership position in the struggle against the threats and consequences of climate change is critical. The main reason for climate change is the buildup of greenhouse gases in the atmosphere.

As observed by USCCSP (2009):

> Consensus in the climate science community is that the global climate is changing, mostly due to humankind's increased emissions of greenhouse gases such as carbon dioxide, methane, and nitrous oxide, from burning of fossil fuels and land-use change (measurements show a 25 percent increase in the last century). Warming of the climate system is unequivocal, but the effects of climate change are highly variable across regions and difficult to predict with high confidence based on limited observations over time and space (p. 10).

It has been scientifically proven that the leading causes of the increased greenhouse gases in the Earth's atmosphere are human activities in exploiting natural resources and energy production and consumption. NASA (2016) has provided a stark warning of consequences of climate change such as "temperatures will continue to rise; the frost-free season will lengthen; changes in precipitation patterns; more droughts and heatwaves; hurricanes will become stronger and more intense; sea level will rise 1–4 ft by 2100, and the Arctic is likely to become ice-free."

The analysis of the localization of climate change impact is the initial step to build resilience in the community and prepare cities to strengthen disaster recovery and make communities safer. Climate resilience is an adaptation to the adverse environmental changes and the economic and social shocks exacerbated by climate change and mitigation of the current negative effects of energy production. For a resilient community, all essential supporting systems need to be adaptable, flexible, and vigilant. Therefore, climate resilience offers a dynamic approach to the endurance of humans and nature and allows communities to react to disasters.

The dynamic nature of climate change influences all sectors, affecting the entire system of industries, communities, and layers of society. Organizations deploy resilience and sustainability strategic planning to confront governance, economic, environmental, and social challenges stemming from climate change and the recent pandemic in their organizational and leadership capacity. Both climate adaptation and climate mitigation are included in the before-mentioned plans. Most thriving sustainability and resilience plans cover all aspects of organizational functions. Recognition of climate change threats is a conventional component of local planning to create more viable and resilient communities. As Fig. 1.4 indicates, mitigation and adaptation activities are interlinked, and both are necessary for organizational resilience and sustainability strategic planning.

All facets of the built environment need to be incorporated into the resilience planning. Certain mitigation strategies may fall into adaptation strategies and vice versa. The preparation includes updates to emergency management and disaster preparedness plans to protect climate-

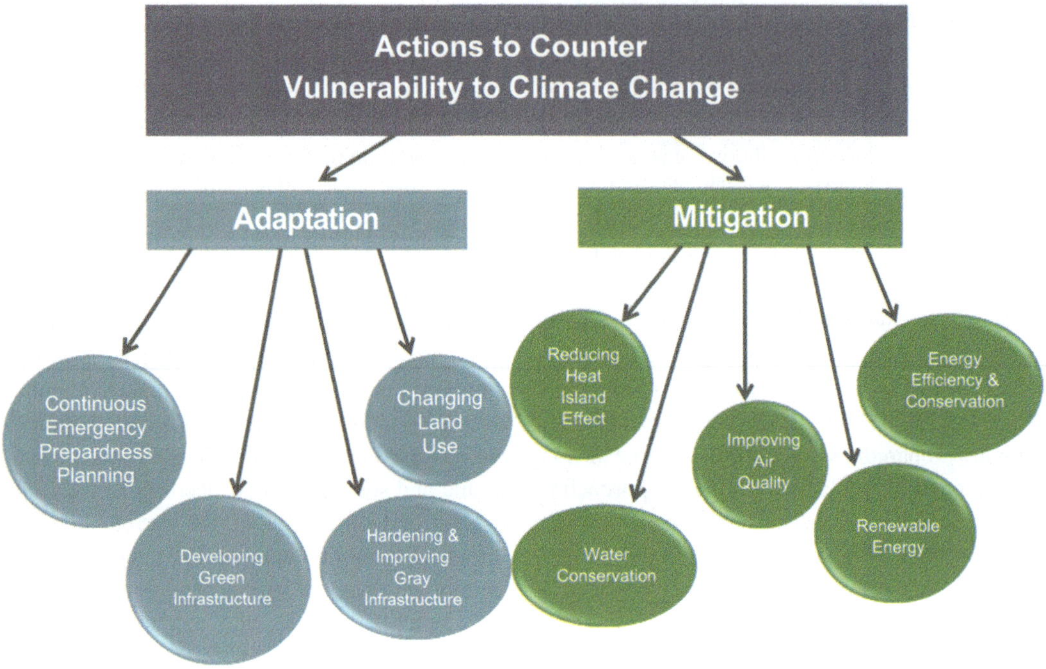

Fig. 1.4 Climate change adaptation and mitigation strategies

vulnerable citizens, developing green infrastructure while hardening and expanding gray infrastructure, and anticipating events not yet experienced. Municipal planners consider additional protective measures to reduce heat island effects, improve air quality, reduce water usage, increase renewable energy production, and reduce energy consumption in operations.

For example, Brownsword et al. (2004) developed a consistent model for energy supply and demand in city settings. The research adds a new dimension to a methodology of evaluating sustainable energy and corresponding to sustainable cities. It offers insight and practical implications for organizations for utilizing sustainable energy and sustainability planning. Other cities have used energy efficiency policies to reduce energy consumption in residential buildings, like in the case of Gainsville, Florida (Douthat et al. 2020).

Resilient Community and Organizations Spotlight: Strategic Sustainability and Resilience Planning in the City of Tallahassee

The City of Tallahassee is the capital of Florida and adopted one of the most comprehensive strategies for sustainability and resilience planning with over 196,000 people living in the urban area, according to the most recent United State Census (2021) data. The City has one of the most advanced strategic resilience and sustainability plans. Its pioneering advancement of environmentally sustainable policies expanded into economic, social, and governance priorities. Tallahassee has a Gold designation in Leadership in Energy and Environmental Design (LEED), a green building certification program recognizing the number of energy-efficient buildings in the city.

The City of Tallahassee's Sustainability and Resilience Division is tasked with community planning, program development, and performance tracking toward a more sustainable and resilient future (City of Tallahassee 2021). The system-wide approach includes city departments, elected and appointed officials, and community-wide partners.

Under its sustainability objectives, the City includes the following:

- 100% renewable energy target;
- Micro mobility investments to enhance transportation infrastructure;
- Neighborhood outreach programs to promote energy efficiency and resource management;
- City farm TLH project, coupled with farm training to promote urban farming;
- Community gardening is a streamlined government program to encourage residents to initiate community gardens;
- Urban forestry is a project adopted to increase the number of trees in an urban setting within the city's borders, aimed at reducing heat island index;
- Brownfield projects are utilized for sustainable economic development, equality, and growth;
- Bio-fuel collection program used to create clean biodiesel;
- Energy efficiency initiatives by the City of Tallahassee Public Utilities aimed at encouraging residents to use energy-efficient appliances and reduce energy consumption. The comprehensive list of initiatives includes budget billing, free home energy audits, neighborhood REACH, and solar net metering (City of Tallahassee n.d.);
- Tallahassee Solar is a program aimed at providing City utility customers with solar panels without upfront costs;
- Think Before You Throw and Think About Personal Pollution are the two programs to reduce litter and pollution by the city's residents (City of Tallahassee 2021; T.A.P.P. n.d.).

Significant Commitment to Resilience Planning

The City's approach to community resilience deserves a closer review. The city's community resilience plan focuses on the more significant implications for the entire community. The City of Tallahassee (2019) defines resilience as "making a city stronger, in both good times and bad, for the benefit of all its citizens, beginning with the most vulnerable" (p. 4). Community collaboration on sustainability and resilience from energy efficiency, renewable energy, pollution reduction, home improvements, food sustainability, the urban tree canopy is evidence of the importance of partnerships to achieve successful sustainability and resilience-related outcomes. The City of Tallahassee (2019) resilience plan identifies four distinct goals:

1. Public Safety and Preparedness, supporting governance mechanisms into preparing households, individuals, and partnering organizations for stressors and shocks.
2. Hazard Mitigation and Climate Adaptation, with a focus on future threats and environmental protection.
3. Equity and Social Cohesion, with an emphasis on the social aspects of the community.
4. Planning and Integration, an effort to integrate resilience into all organizational and community plans.

Summary

The expected outcomes of this chapter included defining strategic resilience and sustainability planning, and delineating all the elements, systems contributing to robustness and longevity of successful planning process. The resilience and sustainability strategic planning and the impact on governance, economic, environmental, and social bottom line enhances the service delivery efforts for organizations and improves communities. The positive social and economic results from sustainability and resilience planning include reduced greenhouse gas emissions resulting from the production of electricity from traditional sources of energy, reduced reliance on oil imports, the lessened negative effects on the environment, improved governance, better service delivery, the ability to withstand shocks, recover from disasters, prepare for climate change, adapt to stresses, more diverse, equitable, and just community and other societal benefits. The positive governance outcomes include more transparency, answerability, and accountability. The ongoing COVID-19 pandemic has further exacerbated stresses and shocks to organizations and communities. The inclusive, system-wide approach to integrating resilience and sustainability within the organizational and community structures is critical to successful long-term planning.

Cities' and counties' investments in resilience and sustainability planning and implementation have a positive social impact. The reduced investment in energy can free up capital to be invested into other services provided by local governments, such as infrastructure improvements, public safety, education, and public health. Once sustainability and resilience performance is measured, tracked, reported, observed, and then compared to various outcomes, it assists organizations in assessing the positive impact on the overall effectiveness of local government service delivery. All societal issues can be potentially presented using resilience and sustainability as a lens or a conceptual framework. Resilience and sustainability plan becomes an organizational strategic plan and integral part of strategic management. The planning process must be intentional and include all elements and stakeholders of the organization and the community.

Outcomes, Discussions, and Further Considerations

- Discuss the differences between resilience and sustainability planning and strategic planning.
- Define resilience and sustainability planning and the importance of both for local governments and communities.
- Describe the history and evolution of sustainability, and the next phase of sustainability and resilience planning.
- Analyze the critical characteristics of resilience and sustainability efforts and sustainable and resilient organizations.
- Assess the climate change risks to local governments and their effects on operations and service delivery.
- Determine the appropriate levels of climate change preparedness and disaster responses.

References

Agyeman J (2005) Sustainable communities and the challenge of environmental justice. New York University Press, New York

Alibašić H (2017) Measuring the sustainability impact in local governments using the Quadruple Bottom Line. Int J Sustain Policy Pract 13(3):37–45

Alibašić H (2018a) Redesigning organizations for maximum resiliency in dynamic environments. In: Farazmand A (ed) Global encyclopedia of public administration, public policy, and governance. Springer International AG, Zurich. https://doi.org/10.1007/978-3-319-31816-5

Alibašić H (2018b) Sustainability as organizational strategy. In: Farazmand A (ed) Global encyclopedia of public administration, public policy, and governance. Springer International AG, Zurich. https://doi.org/10.1007/978-3-319-31816-5_3433-1

Alibašić H (2018c) Role of corporations in addressing climate change. In: Farazmand A (ed) Global encyclopedia of public administration, public policy, and governance. Springer International Publishing AG, Cham. https://doi.org/10.1007/978-3-319-31816-5_3429-1

Alibašić H (2018d) Paris climate accord sans the USA: the role of local governments in confronting the effects of climate change and increasing resilience. Global Stud J 11(2):33–39. https://doi.org/10.18848/1835-4432/CGP/v11i02/33-39

Alibašić H (2018e) The nexus of sustainability and climate resilience planning: embedding climate resilience policies in local governments. Int J Clim Change Impacts Response 10(2):21–33. https://doi.org/10.18848/1835-7156/CGP/v10i02/21-33

Alibašić H (2018f) Sustainability and resilience planning for local governments: the quadruple bottom line strategy. Springer International Publishing, Berlin

Alibašić H (2020) The administrative and ethical considerations of climate resilience: the politics and consequences of climate change. Public Integrity:1–18. https://doi.org/10.1080/10999922.2020.1838142

Ammons DN, Smith KW, Stenberg CW (2012) The future of local government: will current stresses bring major, permanent changes? State Local Govt Rev 44(1):64S–75S. https://doi.org/10.1177/0160323X12454143

Baumgartner RJ, Rauter R (2017) Strategic perspectives of corporate sustainability management to develop a sustainable organization. J Cleaner Prod 140:81–92

Becker C (2011) Sustainability ethics and sustainability research. Springer Netherlands, Germany

Bendell J, Kearnis K (2005) The political bottom line: the emerging dimension to corporate responsibility

for sustainable development. Bus Strategy Environ 14(6):372–383. https://doi.org/10.1002/bse.439
Benn S, Edwards M, Williams T (2014) Organizational change for corporate sustainability. Routledge, New York
von Bertalanffy L (1950) An outline of general system theory. Br J Philos Sci 1(2):134–165. https://doi.org/10.1093/bjps/I.2.134
von Bertalanffy L (2013) General system theory: foundations, development, applications. Braziller, New York. (Original work published 1969)
Brownsword RA, Fleming PD, Powell JC, Pearsall N (2004) Sustainable cities—modeling urban energy supply and demand. Appl Energy 82:167 180. https://doi.org/10.1016/j.apenergy.2004.10.005
Bryce C, Ring P, Ashby S, Wardman JK (2020) Resilience in the face of uncertainty: early lessons from the COVID-19 pandemic. J Risk Res 23(7–8):880–887. https://doi.org/10.1080/13669877.2020.1756379
Bryson JM (2004) What to do when stakeholders matter: stakeholder identification and analysis techniques. Public Manage Rev 6(1):21–53
Cafer A, Green J, Goreham G (2019) A community resilience framework for community development practitioners building equity and adaptive capacity. Commun Dev 50(2). https://doi.org/10.1080/15575330.2019.1575442
Center for Climate and Energy Solutions (2021) Federal action on climate. Center for Climate and Energy Solutions.. https://www.c2es.org/content/federal-action-on-climate/
City of Tallahassee (2019) Community resilience plan. Adopted July 2019. https://www.talgov.com/uploads/public/documents/publicsafety/resplansm.pdf
City of Tallahassee (2021) Sustainability. https://www.talgov.com/sustainability/sustainability.aspx
City of Tallahassee (n.d.) Products & services for city utility customers. https://www.talgov.com/you/you-products-home-index.aspx
Coyle SJ (2011) The built environments and its supporting systems. In: Coyle SJ (ed) Sustainable and resilient communities: a comprehensive action plan for towns, cities, and regions. Wiley, Hoboken, pp 1–13
Daly HE (1996) Beyond growth: the economics of sustainable development. Beacon Press, Boston
Davis J, Alibašić H, Norris S (2018) Corporate leadership in sustainability: a green ranking performance-based approach to understanding corporate social responsibility (CSR) and positive marketing impact. Creighton J Interdiscip Leadersh 4(1):3–6
Douthat T, Morgan JD, Alibašić H, Grant A (2020) Passive building characteristics, and summertime residential energy use: a spatial analysis of energy efficiency in Gainesville, FL. Build Environ 169:106542
Elkington J (1997) Cannibals with forks: the triple bottom line of 21st century business. Capstone, Oxford
Elkington J (2012a) Enter the triple bottom line. In: Henriques A, Richardson J (eds) The triple bottom line: does it all add up? Assessing the sustainability of business and CSR. Earthscan, London, pp 1–17
Elkington J (2012b) The zeronauts: breaking the sustainability barrier. Routledge, Abingdon
Elster J, Hylland A (eds) (1989) Foundations of social choice theory. Cambridge University Press, New York
Federal Resilience Policy. Center for Climate and Energy Solutions (2021). https://www.c2es.org/content/federal-resilience-policy/
Fernández-Prados JS, Lozano-Díaz A, Muyor-Rodríguez J (2021) Factors explaining social resilience against COVID-19: the case of Spain. Eur Soc 23(Suppl 1):S111–S121. https://doi.org/10.1080/14616696.2020.1818113
Fitzgerald J (2010) Emerald cities: urban sustainability and economic development. Oxford University Press, New York
Fletcher C (2013) Climate change: what the science tells us. Wiley, Hoboken
Gaertner W (2009) A primer in social choice theory, revised edn. Oxford University Press, New York
Galea C (ed) (2004) Teaching business sustainability: from theory to practice. Greenleaf Publishing, Sheffield
Girardet H (2006) Creating sustainable cities. Schumacher briefings, Devon
Gray R, Milne M (2002) Sustainability reporting: who's kidding whom? Chartered Accountants J New Zealand 81(6):66–70
Hardjono T, Van Marewijk MV, de Klein PD (2004) Introduction to the European Corporate Sustainability Framework (ECSF). J Bus Ethics 55(2):99–113. https://doi.org/10.1007/s10551-004-1894-x
Heal GM (1998) Valuing the future: economic theory and sustainability. Economics for a sustainable earth series. Columbia University Press, New York
IPCC (2014) Climate change 2014: synthesis report. In: Core Writing Team, Pachauri RK, Meyer LA (eds) Contribution of working groups I, II and III to the fifth assessment report of the intergovernmental panel on climate change. IPCC, Geneva, 151 pp
Koch H, Franco ZE, O'Sullivan T, DeFino MC, Ahmed S (2016) Community views of the federal emergency management agency's "whole community" strategy in a complex US city: re-envisioning societal resilience. Technol Forecast Social Change 121:31–38. https://doi.org/10.1016/j.techfore.2016.12.009
Loorbach D (2007) Governance for sustainability. Sustainability 3(2):1–4. https://doi.org/10.1080/15487733.2007.11907996
Magis K (2010) Community resilience: an indicator of social sustainability. Soc Nat Resour 23(5):410–416. https://doi.org/10.1080/08941920903305674
Martin LL, Levey R, Cawley J (2012) The "new normal" for local government. State Local Govt Rev 44(1):17S–28S. https://doi.org/10.1177/0160323X12440103
Matarrita-Cascante D, Trejos B, Quin H, Joo D, Debner S (2017) Conceptualizing community resilience: revisiting conceptual distinctions. Commun Dev 48(1):105–123. https://doi.org/10.1080/15575330.2016.1248458
Mufson S (2019) More U.S. businesses making changes in response to climate concerns. The

Washington Post. https://www.washingtonpost.com/climate-environment/more-us-businesses-making-changes-in-response-climate-concerns/2019/06/10/a30c86ac-8944-11e9-98c1-e945ae5db8fb_story.html

Murray T (2019) The businesses that are—and are not—leading on climate change. Forbes. https://www.forbes.com/sites/edfenergyexchange/2019/11/08/the-businesses-that-are%2D%2Dand-are-not%2D%2Dleading-on-climate-change/?sh=2abfd777aa15

Ostrom E (1990) Governing the commons: the evolution of institutions for collective action. Cambridge University Press, Cambridge

Pacthod D, Pinner D (2021) Commentary: time is running out for business leaders who don't have a 'net zero' strategy. Fortune. https://fortune.com/2021/04/22/corporate-net-zero-carbon-emissions-plans-climate-change-business-strategy/

Patton MQ (2002) Qualitative research and evaluation methods, 3rd edn. Sage Publications, Thousand Oaks

Saha D (2009) Factors influencing local government sustainability efforts. State Local Govt Rev 41:39–48. https://doi.org/10.1177/0160323X0904100105

Savitz AW, Weber K (2006) The triple bottom line: how today's best-run companies are achieving economic, social, and environmental success- and how you can too. Wiley, San Francisco

Senge PM (2010) The fifth discipline: the art & practice of the learning organization. Crown, New York

Slavin MI (2011) The rise of the urban sustainability movement in America. Creating the green metropolis. In: Slavin MI (ed) The rise of the urban sustainability movement in America. Island Press, Washington, DC

Solow RM (1992) An almost practical step toward sustainability: an invited lecture on the occasion of the Fortieth Anniversary of Resources for the Future. Resources for the Future. Island Press, Washington, DC

Son C, Sasangohar F, Peres CS (2017) Redefining and measuring resilience in emergency management systems. In: Proceedings of the human factors and ergonomics society annual meeting, vol 61(1). https://doi.org/10.1177/1541931213601899

State, Local, and Tribal Leaders Task Force on Climate Preparedness and Resilience, The (Task Force) (2014) Recommendations to the President. https://obamawhitehouse.archives.gov/sites/default/files/docs/task_force_report_0.pdf

Stubbs W, Cocklin C (2008) Conceptualizing a "sustainability business model". Org Environ 21(2):103–127. https://doi.org/10.1177/1086026608318042

The National Aeronautics and Space Administration (NASA) (2016) Global climate change. http://climate.nasa.gov/effects/

The National Aeronautics and Space Administration (NASA) (2021) Climate change evidence: how do we know? NASA. https://climate.nasa.gov/evidence/

The Senate and House of Representatives of the United States of America in Congress (The Congress) (1969) The National Environmental Policy Act of 1969, as amended. https://www.energy.gov/sites/default/files/nepapub/nepa_documents/RedDont/Req-NEPA.pdf

The U.S. Climate Change Science Program and the Subcommittee on Global Change Research (USCCSP) (2009) Synthesis and assessment product 4.1 report by the U.S. climate change science program and the subcommittee on global change research. http://www.globalchange.gov/sites/globalchange/files/sap4-1-final-report-all.pdf

The U.S. Global Change Research Program (USGCRP) (2014) The 2014 national climate assessment. http://nca2014.globalchange.gov/downloads

The United Nations Framework Convention on Climate Change (UNFCCC) (2014) Kyoto protocol. http://unfccc.int/kyoto_protocol/items/2830.php

The White House (2021a) Paris climate accord. https://www.whitehouse.gov/briefing-room/statements-releases/2021/01/20/paris-climate-agreement/

The White House (2021b) FACT SHEET: President Biden takes executive actions to tackle the climate crisis at home and abroad, create jobs, and restore scientific integrity across federal government. https://www.whitehouse.gov/briefing-room/statements-releases/2021/01/27/fact-sheet-president-biden-takes-executive-actions-to-tackle-the-climate-crisis-at-home-and-abroad-create-jobs-and-restore-scientific-integrity-across-federal-government/

The White House (2021c) Press briefing by Press Secretary Jen Psaki, Special Presidential Envoy for Climate John Kerry, and National Climate Advisor Gina McCarthy, January 27, 2021. https://www.whitehouse.gov/briefing-room/press-briefings/2021/01/27/press-briefing-by-press-secretary-jen-psaki-special-presidential-envoy-for-climate-john-kerry-and-national-climate-advisor-gina-mccarthy-january-27-2021/

Think About Personal Pollution (TAPP) (n.d.) Think about personal pollution (TAPP). https://tappwater.org/

Tveiten CK, Albrechtsen E, Wærø I, Wahl AM (2012) Building resilience into emergency management. Saf Sci 50:1960–1966. https://doi.org/10.1016/j.ssci.2012.03.001

United Nations Conference on Environment & Development (1992) Agenda 21. https://sustainabledevelopment.un.org/content/documents/Agenda21.pdf

United Nations Framework Convention on Climate Change (UNFCCC) (2015a) Adoption of the Paris agreement. United Nations Framework Convention on Climate Change. https://unfccc.int/resource/docs/2015/cop21/eng/l09r01.pdf

United Nations Framework Convention on Climate Change (UNFCCC) (2015b) Paris agreement. https://unfccc.int/sites/default/files/english_paris_agreement.pdf

United Nations World Commission on Environment and Development (1987) Our common future. Oxford University Press, Oxford. http://www.un-documents.net/wced-ocf.htm

United State Census Bureau (2021) QuickFacts, Tallahassee city, Florida. https://www.census.gov/quickfacts/tallahasseecityflorida

Visseren-Hamakers IJ (2018a) Integrative governance: the relationships between governance instruments taking center stage. Environ Plan C Polit Space 36(8):1341–1354

Visseren-Hamakers IJ (2018b) A framework for analyzing and practicing integrative governance: the case of global animal and conservation governance. Environ Plann C Polit Space 36(8):1391–1414

Waddock S (2009) Leading corporate citizens: vision, values, and value added, 3rd edn. McGraw-Hill Irwin, New York

Watts J (2021a) Climate scientists shocked by scale of floods in Germany. The Guardian. https://www.theguardian.com

Watts J (2021b) What is causing the floods in Europe. The Guardian. https://www.theguardian.com

Werbach A (2009) Strategy for sustainability: a business manifesto. Harvard Business Press, Boston

Outlining the Strategic Resilience and Sustainability Planning Process for Communities and Organizations

2

"Parties hereby establish the global goal on adaptation of enchaining adaptive capacity, strengthening resilience and reducing vulnerability to climate change, with a view to contributing to sustainable development. and ensuring an adequate adaptation response in the context of the temperature goal ..."

Article 7 of Paris Climate Agreement (UNFCCC 2015)

Overview

The second chapter explains the appropriate processes, steps, and resources in strategic resilience and sustainability planning for organizations and communities. It answers the principal queries related to strategic plans, long-term goals, vision, values, and core mission of an organization pursuing sustainability and resilience planning. The chapter identifies the external and internal dynamics of organizational commitment to strategic resilience and sustainability planning, threats, and strategic opportunities. It evaluates them within the organization's budgetary and financial circumstances. Chapter two maps out the strategic resilience and sustainability planning process for organizations and communities, determining the appropriate steps to take at each planning level. The strategic resilience and sustainability planning process includes the environmental scan, charting the current conditions and available internal resources, the organization's capacity, and analyzing opportunities in the community, region, state, and the nation. It expands on the processes for organizations and communities and determines the practical steps at each level of strategic resilience and strategic sustainability planning.

Keywords

Internal and external environmental evaluation · Current conditions · Internal resources · Accountability · Transparency · Answerability · Accessibility · Diversity · Strategic plan · Sustainability plan · Resilience plan · Charting the process · External pressure · Built environment · Localization of climate change planning systems · 5Is · Climate resilience report

Key Questions

The second chapter answers the underlying assumptions and questions:

- Does an organization have a strategic plan, and what strategies are currently in place that

H. Alibašić, *Strategic Resilience and Sustainability Planning*, Sustainable Development Goals Series, https://doi.org/10.1007/978-3-030-91159-1_2

may be viewed as resilience- and sustainability-related?
- What are an organization's long-term goals, vision, values, and core mission pursuing resilience and sustainability?
- What resilience and sustainability program does the organization have in place, and how are they defined?
- What are the substantive resilience and sustainability efforts in the organization?
- Does the organization have a plan to reduce greenhouse gas emissions or to minimize the impact of climate change?
- Does the organization have a plan for recycling, energy efficiency, renewable energy, and waste minimization?
- What is the current budgetary and financial situation in an organization?
- What are the available internal resources, and what is the organization's capacity?
- How does an organization address its long-term planning; strategies for economic growth, social justice, and equity; environmental protection; good governance; disaster preparedness and resilience?

Introduction

Continuing from the first chapter, this chapter outlines and defines processes in determining the appropriate steps and resources for organizations' resilience and sustainability strategic planning. The chapter is intended to answer the inquiries associated with an existing plan, long-term goals, vision, values, and core mission of an organization pursuing resilience and sustainability planning initiatives. It enables the readers to recognize the external and internal dynamics of organizational commitment to resilience and sustainability, threats, and opportunities and to evaluate them in the context of budgetary and financial circumstances surrounding the organization. The imperative posed by climate change and pandemics increases the need for effective resilience and sustainability planning.

This chapter charts the resilience and sustainability planning process for organizations and communities, determining the appropriate steps to be taken at each sustainability and resilience planning level. The planning process includes the environmental scan, mapping out the current conditions and available internal resources, the capacity of the organization, and an analysis of opportunities that may exist in the community, region, state, and nation. The chapter aims to provide organizations with tools to recognize available resources, coupled with challenges and opportunities. It also enables the leadership to evaluate existing resilience and sustainability activities within organizations.

Organizational planning starts with vision and mission statements, a review of current conditions and existing sustainability efforts, long-term goals, and the budgetary and financial condition of the organization. Various organizations, including local governments, have resilience and sustainability programs through recycling, composting, energy efficiency, waste management, bike lanes, biking programs, watershed protection, renewable energy, and tree planting programs. However, organizations whose administrators and managers do not measure, track, observe, and report corresponding outcomes and results may not fully benefit from resilience and sustainability programs.

A proposed method to resilience and sustainability planning is to apply the Quadruple Bottom Line strategy and measure governance, economic, environmental, and social targets and outcomes. Additionally, facing climate change threats and a recent pandemic, local governments, such as counties, townships, villages, and cities, develop climate action plans, emergency preparedness strategies to account for climate change threats and pandemics, enhanced disaster preparedness strategy, emergency management resilience plans, and climate mitigation and adaptation strategies. A thorough review of the existing plans is warranted, including the comprehensive plan, master plan, traffic safety plans, emergency preparedness plans, hazard mitigation plans, and other relevant programs, policies, and procedures, including the cities' budget and fiscal policies.

Resilience and Sustainability Plan and Quadruple Bottom Line (QBL)

Aligning the resilience and sustainability planning with the fiscal year, budget process, and defining the benchmarks and baseline year is critical to the robust implementation of the resilience and sustainability initiatives. Moreover, aligning benchmark data and targets with the existing plans and policies is essential to the longevity of resilience and sustainability in organizations.

The plan identifies evolving resilience and sustainability priorities, guiding principles, key objectives, and strategies in an era of organizational, community-wide, and climate-related initiatives and transformation efforts. The goals and targets are outlined in the resilience and sustainability plan to equip local government officials to respond to future obstacles and opportunities in a viable and coordinated manner. The strategic resilience and sustainability plan acts as a conduit to various city's plans and expands the Triple Bottom Line framework into a Quadruple Bottom Line (QBL) structure. Quadruple Bottom Line (QBL) functions as the conduit to overarching elements of the plan: governance, economic, environmental, and social. For illustrative purposes, specific objectives of a plan may be placed under one of the four elements of the Quadruple Bottom Lines, and specific targets are categorized under separate objectives: economic and commercial opportunity, prosperous neighborhoods, social equity and fairness, safe community, resilient systems, balanced transportation, resilient infrastructure and assets, fiscal resilience, transparent and accessible, and resilient and sustainable government.

For cities with an existing plan, after establishing a baseline or expanding upon an existing baseline, the revised resilience and sustainability plan constructs upon the previous plans. In the past, these plans relied on the Triple Bottom Line (TBL) framework. However, a Quadruple Bottom Line (QBL) structure allows governance and related efforts to be appropriately inserted into the resilience and sustainability plan. Also, drawing from the city's existing plans and aligning them with its goals, outcomes, and objectives allow for consistency, robustness, continuity, and integrity of operations.

The Quadruple Bottom Line offers the structure for a citywide resilience and sustainability strategic plan with governance, economic, environmental, and social elements. Each specific target is categorized under separate objectives and outcomes, directly connected to the city's budgetary and fiscal plans, as suggested below for illustrative purposes:

1. Good governance: accountability, transparency, accessibility, community input, disaster preparedness, emergency management, and fiscal resilience.
2. Resilient assets, infrastructure, buildings, utilities, energy, and balanced transportation.
3. Resilient economic growth and opportunities.
4. Resilient environment and natural resources.
5. Resilient and prosperous neighborhoods.

Accountability, answerability, transparency, and accessibility are vital to ensuring communities and city governments are more resilient and sustainable. Organizations, citizens, institutions, and businesses make decisions, undertake actions, and implement activities that impact a community's overall resilience and sustainability. It is the responsibility of appointed and elected officials and employees to contribute to implementing resilience and sustainability targets. Furthermore, staff assigned to be the outcome or target champions are responsible for leading and achieving specific resilience and sustainability targets outlined in the plan.

The resilience and sustainability strategic plan enables individual accountability and ownership for each goal for effective results. Organizational resilience and sustainability strategic plans further good governance and improve operational efficiency and social equity in the community. The plans are utilized to address affordable housing opportunities, diversity, and inclusiveness, reduce energy consumption, intensify renewable energy production, minimize waste, support community outreach efforts, standardize the system processes, partner with organizations, companies, and citizens to further spur growth while preserving natural resources and to prepare for climate change and pandemics.

Municipal employees track the progress of each target and create a public record outlining both the plans and actions that address the Quadruple Bottom Line (QBL). The resilience and sustainability progress reports are released to the public annually or biannually. However, cities and counties alone cannot implement all of the strategies outlined in the plans and strive to partner and collaborate with other organizations within the community, region, or at the state and national stage.

Strategic vs. Sustainability Planning in Cities

Certain local governments continue to utilize both the strategic plan and resilience and sustainability plan. An ideal situation would be to use the strategic resilience and sustainability plan, with a set of specific, well-defined Quadruple Bottom Line targets. The comparison of benefits for strategic resilience and sustainability plan in place of strategic plan charts the process and features the benefits of choosing the organization's resilience and sustainability plan. Setting a defined, target-driven plan in place with a set of measurable outcomes enables local government administrators to deal better with ongoing demands for services, achieve long-term objectives, deliver outcomes, and define a long-term vision for the future.

Benefits of a Strategic Resilience and Sustainability Plan

- Contains the progressive effort with specific targets, baseline years, and benchmarks adjusted for annual outcomes.
- Includes continuous events applied to unambiguous programs and policies.
- Allows for planning for disasters, climate change, pandemics, and emergency management.
- Includes a system-wide strategy to energy management, food production, urban tree canopy, transportation, sustainable energy, mobility, economy, environment, equity, traffic, water, sewer, city operations, governances, management, and administration.
- Develops a broader set of dynamic and measurable goals.
- Enables staff involvement and community outreach in a bottom-up approach and with public engagement.
- Measures and reports targets, goals, and outcomes.

Disadvantages of a Strategic Plan

- Static, a one-time event-oriented plan.
- Is limited in scope to management.
- Has a short-term focus and goals.
- Involves top-down approach with no input from staff or the community.
- Does not measure governance, economic, environmental, and social objectives and targets.

While the two types of plans appear similar; in practice, they are different in meeting organizational objectives. A strategic plan relies on a static, linear definition of goals and objectives with a short-term focus. Witcher (2020) posited that strategic management consists of strategic planning and implementation and processes using a leadership top-down approach. On the other hand, the strategic resilience and sustainability plan is multidimensional and meets objectives from a long-term perspective from economic, social, environmental, and governance perspectives. The resilience and sustainability plan offers a more progressive and practical approach to measure results.

By choosing a resilience and sustainability strategic plan, organizations avoid duplication of efforts. Beyond manageable strategic components, well-defined resilience and sustainability planning efforts facilitate consistent service delivery, operations continuity, and long-term goals and objectives. Organizations committed to environmental stewardship, social equity, sustainable economic growth, and good governance combine climate resilience activities and sustainability planning into one. By taking such an approach, local government organizations constructively amalgamate their planning efforts into a single visioning and guiding document, ensuring consistency in planning and reporting. The strategic resilience and sustainability planning process inevitably leads to a more vibrant and resilient community, effective government operations, better preparedness for climate change and pandemic responses. Those outcomes contribute to an overall reduction of greenhouse gas emissions and lower carbon footprint locally, regionally, and globally.

In response to frequent threats exacerbated by climate change and extreme weather, cities undertake activities to counter climate change threats. Boswell et al. (2012) and United Nations Climate Change (2021a, b) viewed climate mitigation planning as strategies to reduce greenhouse gas emissions and the use of climate adaptation as a strategy to bolster resilience. The scope and approach to resilience and sustainability strategic planning include the climate action plan, with a broader view on the community's sustainability, climate adaptation, mitigation, readiness, and preparedness strategies. Climate action strategies and plans are integrated into an existent resilience and sustainability plan.

Resilience and sustainability as a policy and programmatic framework are used by organizations in both the private and the public sectors. With its precision, practicality, and applicability, strategic resilience and sustainability planning leaves a significant impact on organizations, and the practicality and applicability leave a positive impact on organizations and communities. Importantly, existing resilience and sustainability planning enables the municipal administrators to apply the practical elements of Quadruple Bottom Line to municipal operations and service delivery.

Strategic resilience and sustainability planning is an innovative, original, and novel approach organizations employ to create more resilient and sustainable communities. Focusing on organizational effectiveness and efficiency through the successful application of resilience and sustainability enables practical implications for the longevity and robustness of ongoing operations. In their research, Cumo et al. (2012) showed that concrete transformation and changes occur in urban settings and should be taken seriously under consideration. Pierce et al. (2011) studied the connection between resilience and sustainability in 40 urban areas and have concluded a degree of connection between resilience and sustainability in urban areas and the effects on social and governance aspects of communities. Magis (2010) concluded how community resilience is a factor in defining social sustainability. When evaluating urban areas and their negative environmental impact, the exact definitions of an urban area are often under scrutiny. Kennedy and Sgouridis (2011) argued the difficulty of delineating the exact urban boundaries when determining the greenhouse gas emissions impacts.

There are multiple steps involved in charting the resilience sustainability planning in organizations and presenting its auxiliary impact on a community. The initial step to resilience and sustainability planning is defining a mission and a vision statement, aligning them with the long-term organizational objectives, determining the type of activities to be measured and what level, defining stakeholders, and prioritizing areas of responsibility. While it relates to the city as an organization with its operations and resources, the city's resilience and sustainability plan also focus on the broader implication of resilience sustainability in the region and its impact on the community.

A strategic resilience and sustainability plan serves as a guide for organizational actions, activities, and initiatives. The resilience and sustainability plan serves as a long-term strategy, a comprehensive plan relying on climate science and the study of extreme weather patterns, analysis of vulnerabilities, and potential unforeseen threats such as pandemics.

Identifying the Current Conditions and Resources

The external pressures on the built environment are best explained using the Quadruple Bottom Line strategy. Governance factors are considered under the management, administration, and operations of the local government administrators and managers. The demographic changes and trends, the income level of the city population, and socioeconomic movements fall under the social and economic categories. Environmental factors include the quality of water, air, available natural resources, industrial pollution, and other determinants impacting the quality of the environment.

Leadership plays a relevant role, with staff involvement, community engagement, accountability, transparency, fiscal responsibility, answerability, ethics, and public service integrity. Droege (2006) argued the importance of urban areas and cities in addressing the increases in fossil fuel-related emissions. Cities are principal contributors to the overall greenhouse gas emissions in the world. Measures to counteract the effects of greenhouse gas emissions and climate change include energy conservation; waste minimization supported by reusing, repurposing, and recycling of materials; public transit; bicycling; pedestrian-friendly neighborhoods; quality of life; and a cleaner and greener water infrastructure. Seeking resilience and sustainability-driven policies and practices in cities and counties must include proactive measures in reducing greenhouse gas emissions.

Moreover, Lindfield (2010) observed the negative effects of globalization on cities and service delivery. Local governments in the United States and Canada undertook initiatives to heed the loss of manufacturing jobs amid a shift to new industries and outsourcing of production, without adequate infrastructure investments, and lack of support for education, among many challenges communities have been facing in recent years. In essence, this shift in economic growth created new difficulties for cities. At the same time, some metropolitan cities became centers of economic activities, which increased the adverse environmental impacts on populations.

Resilience and sustainability strategic plans are used as planning and strategy tools and policies adopted on local, national, and international levels for communities and organizations to successfully combat increasing threats and address risks of unsustainable planning policies and practices. Communities are continually seeking strategies to promote the integrity of the natural environment, including energy use decrease, climate protection, improved environmental quality, and natural systems, sustainable land use, urban form, and transportation. By positioning resilience and sustainability planning through the Quadruple Bottom Line, organizations adequately and appropriately address ongoing and unforeseen threats, including disasters caused by climate change and pandemics.

Furthermore, municipalities embrace opportunities expressed in emergent and best practices, such as renewable energy production, electric vehicles and charging stations infrastructure investments, tree planting in urban settings, increasing the availability of affordable housing options, and addressing social equity and other issues (Fig. 2.1). Resilience and sustainability strategic planning empowers organizations to use a multifaceted, cross-sectoral approach for organizations' betterment and operational efficiency. The primary drivers for successful resilience and sustainability programs are the necessities to adapt to the governance, economic, environmental, and societal conditions surrounding organizations. Thorough sustainability and resilience planning process enable organizations to create a transformational culture, allowing staff to embrace the mission and values of everyone for an organization. Moreover, resilience and sustainability planning enables administrators to integrate all elements of the organizational system holistically.

The planning requires a detailed outline and analysis of the organizations' current conditions, including a thorough review of risks and opportunities in the environment. Developing an accurate resilience and sustainability strategic plan goals and objectives should, at a minimum, include a vision statement and long-term and short-term goals and identify potential targets that the city

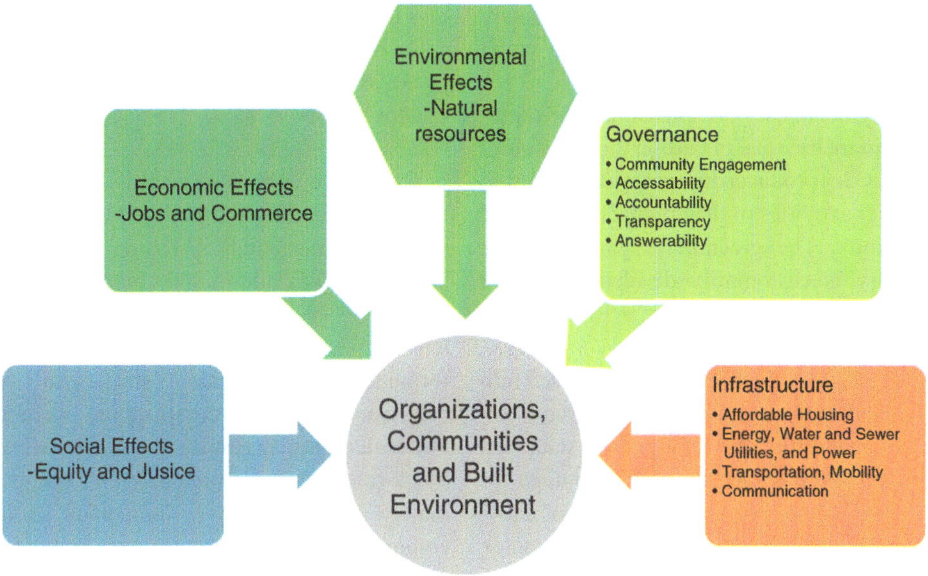

Fig. 2.1 The quadruple bottom line and external pressures in built environment

could adopt. For example, various sustainability planning strategies involve conducting assessments, creating indicators, writing a plan, implementing initiatives, and reporting outcomes.

However, more in-depth strategies and steps involved in the preliminary design of the strategic resilience and sustainability plan should, at a minimum, include the following steps in the process:

1. The first step is a detailed analysis of the vision, mission, and goals, with broad objectives from long- and short-term planning perspectives. Other documents such as a comprehensive plan, master plan, traffic safety plans, parks and recreation plans, emergency preparedness plans, hazard mitigation plans, strategic plans, climate action plans, existing vulnerability assessments, and other reports must be reviewed for alignment.
   (a) An examination and an analysis of the city's relevant documents, the past and current resilience and sustainability-related practices and policies are critical for effective resilience and sustainability strategic planning to frame preliminary recommendations for a plan.
   (b) Interviews with the elected officials and appointed officials need to be included in this process.
2. The second step includes a review of the budget, fiscal conditions, economic growth trends, and social demographics, including population trends, development patterns, and commercial and residential housing demands. The step involves a thorough understanding of the current budget. A city may have a 100% renewable energy target. However, city leadership considers any associated costs with a conversion to cleaner energy sources. The review incorporates the cost of vehicles, equipment, machinery, fuel, and electricity and natural gas use in buildings and operations.
3. A thorough review of the existing energy use by the entire organization, including energy consumption and, in the case where cities own a power utility, energy production. Developing a detailed review of greenhouse gas emissions and defining the outline for the emission target reductions as a result of energy use and energy production. The purpose of the greenhouse gas emissions inventories is to provide a baseline against a particular benchmark or

base year so that staff may more accurately measure progress toward reducing their emissions in different initiatives.

The greenhouse gas emissions are represented in metric tons of carbon dioxide equivalent (CO2e) produced by energy consumption and other activities of the organization and the community. The greenhouse gas emissions inventory is customarily divided into three subcategories: direct city's operations, employees' caused emissions and community-wide emissions. Each category is divided into direct and indirect emissions, with the baseline year and the year from which data was collected. With a myriad of data sources for energy use, demographics, and types of emissions, it is a time-consuming and complex process for cities to conduct the carbon footprint inventory. A more in-depth explanation for greenhouse gas emission or carbon footprint inventory is provided in the Tools and Resources Book Chap. 7.

4. Existing human resources; administrative staff and part-time employees; community-wide resources, nonprofit, health, and human services; private sector partners, and community facilities.
5. A review of existing infrastructure, infrastructure and capital projects plan, including transportation, mobility, roads, sewer, water utilities and facilities, power, and green infrastructure.
6. A review of environmental programs, water and air quality, ozone day programs, waste management, recycling efforts, river, and waterways clean up, and other practices, programs that encourage environmental practices.
7. In the preparatory stage of the resilience and sustainability plan development, an examination, and an analysis of the governance is a critical planning component. Additionally, administrators review the local governments' public engagement, website information, media announcements, social media engagement, communication strategies, neighborhood participation programs, partnership opportunities, and community networks.
8. Finally, a review of existing reporting mechanisms, any software used by departments, and collaborative tools and techniques to aid in reporting outcomes.

Resilience and sustainability planning is a method in identifying the organization's current state, the dimension of its future, and the destination the organizations and communities aim to reach. This examination stage encourages consistent decision-making, communication, and performance assessments and can create cohesiveness with the organization. In embracing these elements, organizations create a building block to enhance the organization's outputs ultimately and improve outcomes for the communities.

Resilience and Sustainability

"Initiate-Implement-Innovate-Inspect-Improve" (5Is) Loop

Resilience and sustainability planning management is a constant process that combines strategic planning and leadership with other management processes. Alibašić (2017) described sustainability "as a set of effective and efficient actions taken by an organization, through good governance, to ensure the economic stability, growth, and financial success, with the most positive societal outcome and the least negative environmental impact" (p. 37). Likewise, Alibašić (2018a, b, c, d) argued that resilience planning is an ethical and leadership imperative to combat the threats of climate change and prepare organizations for maximum resilience to withstand shocks. As such, in resilience and resilience strategic planning, organizations must take into consideration all four components and integrate them effectively into their long-term goals and objectives. The measurement of resilience and sustainability effectiveness through regular reporting mechanisms allows organizations to track progress and enables them to use measures as a tool for refinements and improvements.

The Initiate-Implement-Innovate-Inspect-Improve (5Is) loop approach to resilience and sustainability improvement is a closed-loop cycle for planning, implementing, feedback, and reapplying and refining processes. Resilience and sustainability planning has a continuous loop function, intended for incessant advancement and upgrade. Initiating programs or policies is the first step in the improvement loop. The second step is the implementation phase, while the third allows for innovation to ensure goals are met. The fourth phase is the process of measuring, tracking, observing, and inspecting targets, results, and outcomes for potential improvements. The final stage in the process before returning to the initial step involves improvements and enhancement to the process.

It is imperative for an organizational leadership team to use resilience and sustainability planning in response to the rapid changes occurring in the surrounding environment. Resilience and sustainability planning must be viewed as a dynamic, ever-changing, all-hands-on-deck approach to every level of organizational governance, employed and deployed with maximum resilience in mind.

The resilience and sustainability planning process allows organizations to substantially address critical planning and avoid a crisis in a situation required on any scale. Strategic resilience and sustainability plan is a forward-thinking system of methodologies, evaluating and analyzing developing practices, opportunities, or perils to the organization and developing a comprehensive response, taking advantage of the existing internal and external resources. Organizations can respond to threats in a purposeful mode, with resilience and elasticity as ultimate goals.

Resilience and sustainability planning allows the executives and staff to work together and to better enhance the fulfillment of the mission, goals, vision, and meeting of mandates, with endless improvements and the sustained performance. This package of structural definition allows an organization to function effectively.

Similar to strategic planning, resilience and sustainability planning can start with the three what questions: what is the strategic position of the organization now, what does the organization strive to accomplish and achieve, and what resources does the organization have at its disposal to meet the goals and objectives.

Answering specific sustainability "what" questions will conclusively lead to a change and the path forward in mission, vision, and goals. While historical crises, such as financial recessions and resource shortages recur with vigor, new challenges and opportunities, including but not limited to pandemics, climate change, artificial intelligence, social media, and cyber warfare, have emerged to present nascent leaders to contend with challenges of the contemporary, fast-paced, and technology-reliant world.

As local governments face repeated scrutiny, resilience and sustainability initiatives provide administrators with new purpose to guide their organizations to a more predictable future with clear goals and an understanding of purpose. The Quadruple Bottom Line assists organizations in establishing a path forward and a sense of direction for the organization. An essential aspect of sustainability and resilience strategic planning as it relates to public sector organizations is the nurturing of the transformational culture within an organization, which links planning and implementation and continually challenges the notion of the status quo.

The *Initiate-Implement-Innovate-Inspect-Improve* loop applies specifically to public service organizations to foster an atmosphere and organizational culture where all members are focused on carrying out the mission and values of the organization with an understanding of the direction that the organization is headed. The Initiate-Implement-Innovate-Inspect-Improve loop illustrates the significance of direct loop and feedback between the initial resilience and sustainability plan and the implementation, with the return on investment from resilience and sustainability initiatives connecting to the city's budget annually.

The 5Is loop enables the connection of and the assessment of the existing plans, allowing organizations to measure progress using a resilience and sustainability plan. In this case, a resilience and sustainability plan is used as a conduit and a

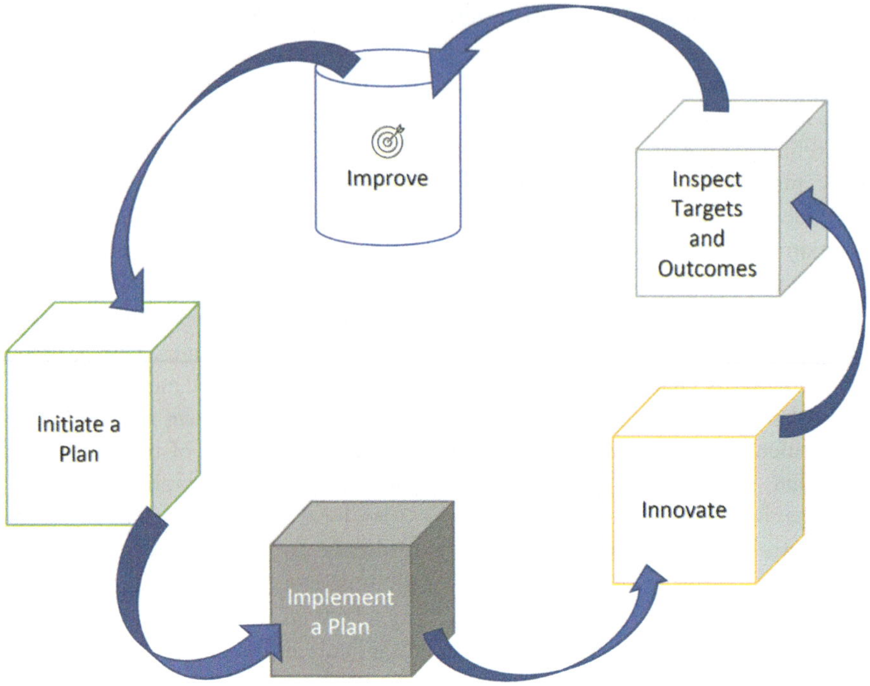

Fig. 2.2 The initiate-implement-innovate-inspect-improve loop

report on the overall outcomes and investments made by the organization. The 5Is loop is efficiently fused with core operations, resulting in savings and process improvements (Fig. 2.2).

Resilient Communities and Organizations Spotlight: Resilient New Orleans, Louisiana

In 2015, the local government in New Orleans in Louisiana released its first resilience strategy, which outlined steps for making and building a more resilient New Orleans. The resilience plan and strategy continued to evolve, and the City's resilience challenge includes the vision for 2050 New Orleans, with recognition of the changing environment, rapidly accelerating climate change, and the need to adapt to uncertainties (City of New Orleans 2021). According to the most recent US Census Data (2021), the city of New Orleans has a population of over 390,000.

The resilience and sustainability planning has been a lengthy process. Among the steps undertaken by the city of New Orleans was the development of the city of New Orleans Carbon Footprint Report. The report built on the previous carbon footprint document, including the 2001 report, with the baseline year of 1998. Importantly, the city recognized the need to reset the carbon footprint data after the devastating effects of Hurricanes Katrina and Rita in 2005. As noted in the 2009 report, climate change and projected sea-level rise continued to pose significant threats to the city unless current rates of greenhouse gas emissions are drastically curbed and reduced (City of New Orleans 2009).

New Orleans' leaders recognized the disproportionate impact, vulnerabilities of the community, and the effects climate change has on the city. The report recognized the impacts and threats of increased frequency and the intensity of hurricanes, including the ultimate negative consequences to the city and the region, including "higher prices and shortages of basic goods, such as food and energy; increased public expenditures on relief and rebuilding due to extreme weather events; a higher susceptibility to flooding; and a higher rate of infectious diseases and heat-related illnesses and deaths" (City of New Orleans 2009, p. 4).

Cities with unique threats must learn to adapt and mitigate the impact of their operations on the environment and society. The consequences of not doing anything to reduce the carbon footprint are dire. In New Orleans alone, if "no actions were taken to reduce GHG emissions from the date the data were collected, the City of New Orleans would produce 6,451,399 tons CO2e in the year 2030, a 36.5% increase from 2007" (City of New Orleans 2009). One of the adaptation and mitigation activities included in the New Orleans master plan is the rehabilitation and enlargement of the city's urban forest with a specific benchmark for a percentage of the citywide tree canopy by a certain year.

Given the extent of the damage to the trees from hurricanes, the urban tree canopy target addressed the adaptation and mitigation strategies. Besides, the city encourages and promotes tree planting, preservation on both the public and privately owned properties. Because of the vulnerabilities to storms and loss of land area, New Orleans is particularly interested in establishing resilient and sustainable stormwater management practices and protecting environmentally sensitive areas, such as wetlands, from adverse impacts to enhance the city's water-storage capacity during storms and increase protection against storm surges.

In recent years, and with the development of the city's resilient strategy, there is further recognition of the unending impact of climate change and persistent threats that exist in the coastal cities. In the Resilient New Orleans report, in addition to the resilient strategies for regional transportation, promoting sustainability as a growth strategy, reducing redundancy and improving reliability of the power supply, integrating resilience into the decision-making process, and investing in pre-disaster planning and post-disaster recovery, the city focuses on developing strategies and planning for resilience at the neighborhood and business district level (City of New Orleans 2017).

In looking at all the relevant aspects of resilience, the city also emphasized the issue of equity as an essential component to resilience: "Even as we look to the future, we cannot ignore past injustices. Racial inequity is present in every facet of our society—employment and income, education and health, violence and justice, housing and social mobility. To advance as a city, we must confront this reality collectively and seek meaningful ways to address its effects on our institutions, communities, and families. With a strategy that prioritizes racial equity, we will be stronger as a society and more capable of responding to adversity" (City of New Orleans 2017, p. 11).

Corporations oftentimes overlooked social and governance issues in strategic resilience and sustainability planning. Similarly, local governments often fail to observe and include social aspects of resilience and sustainability in planning. Including all elements in the plan strengthens the organizational vision, mission, and goals. Key to the success of an organization's attempt to resilience and sustainability planning is its leadership ability to express its vision and goals for such preparation publicly and to engage the public in seeking support. Integrating steps to increase resilience into the city's plans is imperative to the accomplishment and implementation of such plans. Organizations present a consistent and transparent reporting mechanism to the public and share results while creating opportunities for participation and involvement. Since adopting its resilient strategy, the city of New Orleans intentionally implemented programs and projects to enhance the community's resilience and sustainability.

Local Resilience and Sustainability Planning Vision

As cities and counties complete their preliminary review and examination of the resilience and sustainability inventory, current activities, programs, projects, opportunities, and threats, they proceed with defining a vision for resilience and sustainability. The most critical elements of resilience

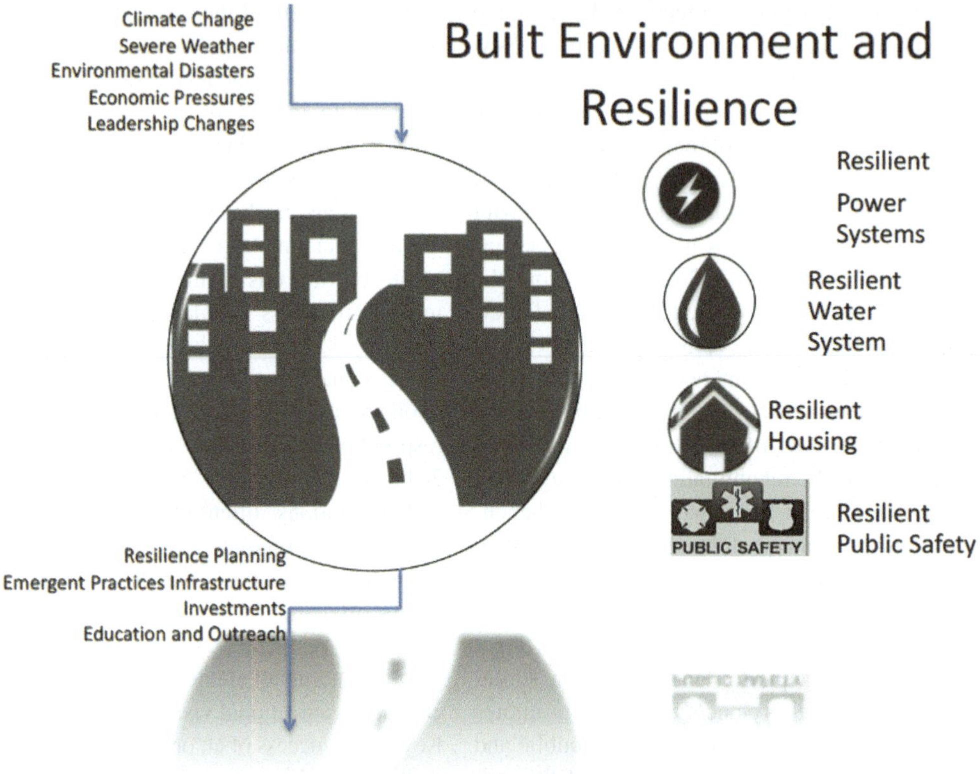

•

Fig. 2.3 Resilience and built environment

and sustainability planning in the built environment include sustainable energy, water wastewater, stormwater, natural environment, mobility, food production, solid waste, economic development, stakeholder engagement, public outreach, and education, as discussed in Alibašić (2018a, b, c, d, e, f). The initial identification of stakeholders and public engagement activities are of utmost importance to strategic resilience and sustainability planning.

The planning process is comprehensive and expansive, and it ensures that organizations and communities continue to operate, function, and thrive before, during and after experiencing a disaster. Cities outline resilience and sustainability planning while accounting for all the elements of the Quadruple Bottom Line. Some of the elements considered in the resilience and sustainability visioning process within a built environment include the following, as presented in Fig. 2.3:

1. *Good governance, accountability, answerability*
   (a) Ensure fiscal responsibility, accountability, and responsiveness.
   (b) Engage the public and educate them about all the aspects of sustainability and resilience and seek constant feedback.
   (c) Relate and report to the public all the progress and failures.
   (d) Engage staff and community organizations, neighborhood and business districts, nonprofits, universities, colleges, schools, and health institutions. It may include regional collaboration or policy labs involving cities and universities as discussed in Alibašić and Crawley (2020).

(e) Support sustainable procurement policies as suggested by Alibašić (2020)

2. *Protection of natural resources and the environment*
   (a) Adopt aggressive water conservation and waterways protection measures.
   (b) Reduce direct discharges to rivers, tributaries, and oceans.
   (c) Promote pollution prevention and reduce toxic chemicals entering waterways, including pharmaceuticals.
3. *Greenhouse gas emissions reduction and clean energy economy*
   (a) Reduce dependence on carbon-based sources of energy.
   (b) Promote renewable energy generation, from wind and solar and geothermal for heating and cooling.
   (c) Improve the efficiency of transmission and hardening of the power grid, including burying power lines to prevent damage during storms.
   (d) Reduce energy consumption and waste.
   (e) Minimize waste and increase composting, waste reduction, repurposing, and recycling.
   (f) Promote efficient and low-carbon transportation alternatives and invest in public transit.
   (g) Support green job creation, research, and innovation, small- and medium-sized businesses.
   (h) Promote eco-friendly tourism and the local economy.
4. *Resilient built environment*
   (a) Adopt low-impact, high-density urban planning with walkable communities.
   (b) Require energy-efficient, green building planning and design.
   (c) Plan for green infrastructure to reduce stormwater, including green roofs, permeable roads, and other measures to reduce water runoff.
   (d) Design and invest in dynamic public transit.
   (e) Invest in green infrastructure.
   (f) Provide quality public safety, supporting first responders and public health agencies.
   (g) Create opportunities for affordable housing and investment in local businesses and neighborhood districts.

Practical Strategy: Climate Resilience and Sustainability Planning Process

Communities are already experiencing negative climate change effects in multiple ways that may include:

- Extreme heatwaves putting the vulnerable population such as elderly, young, and socially disadvantaged at risk;
- More frequent severe rain events, storms, and snowstorms that stress water, road, sewer, power, and other infrastructure;
- More recurrent and intense hurricanes and tornadoes, leading to catastrophic damages and loss of life;
- Water shortages during frequent and intense droughts, leading to fires with catastrophic consequences;
- Increased smog, fire risk, and air pollution that exacerbate respiratory illnesses and other medical conditions;
- Pests and disease risks;
- Pandemic and virus risks;

Using a system-wide approach to evaluate the climate change effects on the local community and regions, cities develop the climate resilience and sustainability report with the localized climate change impact assessment and vulnerability assessment. Notable aspects of the report must include its focus on localization of climate change impact and a distinct list of suggestions for building resilience in the local community and strengthening disaster recovery and resilience of the local government.

Notable goals of the resilience report are to initiate review and further enhance projects, policies, programs, and planning actions enabling cities to mitigate the effects of climate change, to adapt to its impacts, and to utilize emerging sustainability opportunities (Alibašić 2018f). Local community partners and representatives from the public and private sectors from academic, first responders, regional planning, utilities, businesses, universities all contribute to the quality of the local government climate resilience and sustainability planning for the community and the region.

System Approach and Localization of Climate Science

An essential element of combining resilience and sustainability planning is to take a systematic, holistic approach to planning. Climate change affects all sectors and their combination, with system-wide effects acting from local to regional to national and global stages. For successful climate resilience planning stages, the following aspects of weather patterns must be taken into consideration:

- Projections of the climate change data of temperature and precipitation through the future, coinciding with the city's master planning process and comprehensive plans. The annual and monthly baseline averages of temperatures and precipitation should be compared to baseline data for available decades.

  A sample of the analysis and findings that describe climate change impact in a given area consists of:
- Analyze the average temperature and precipitation increases and decreases by oC and percentages over a time period in the future.
- Predicting the seasonality and when the most substantial increases or decreases in temperature are projected during the seasons.
- Calculating the most significant percentage increases or decreases in precipitation per season.
- Assessing the variability and stability of weather over the period of time to predict where more extreme weather events such as storms producing greater than one an inch of rain in 24 h, increased recurrence of back-to-back days above 90° and 90% humidity, more intense and more frequent hurricanes, tornados, dry seasons leading to fires, and more freeze-thaw cycles in winter and spring may occur.

Reporting Conclusions and Recommendations

The climate resilience report includes conclusions and recommendations in the areas of process improvements as described in a sample of recommendations are below as per Alibašić (2018f).

- Under processes, organizations should use a Quadruple Bottom Line cost-benefit approach in financing and implementing of resilient and sustainability projects.
- Seek to adopt decentralized, sustainable energy management toward distributed energy, energy efficiency, and renewable energy systems.
- Continue to encourage the construction of the green building projects.
- Research and implement climate-resilient street maintenance and construction practices, particularly for materials and physical infrastructure.
- Adopt an urban tree canopy goal and implement a forestry program addressing heat island, air quality, and other resiliency values delivered by a diverse, healthy urban tree canopy.
- For coastal areas and islands, include:

  - Comprehensive beach nourishment plans.
  - Sea-level rise predictions.
  - Vulnerability assessment.

Future Climate Resilience Plans

The climate report is used to directly support and link to diverse aspects and targets of the city's resilience and sustainability plan. It offers an opportunity for regional and state-wide discussions on the effects of climate change but also for exact policies and tools to implement climate preparedness and resilience in communities and regions. Individuals or organizations own and champion climate resilience in the community. The climate resilience reporting serves as an amalgam for similar reports for local and regional governments. Boswell et al. (2012) indicated that climate action plans are the primary policy methods to reduce greenhouse gas emissions. The first step to comprehensive strategic resilience and sustainability planning is a thorough and well-thought-out climate vulnerability assessment.

Table 2.1 features recommendations for the climate resilience assessment, translated into actionable targets used in the strategic resilience and sustainability plan Alibašić (2018f). The process used to embed and coordinate resilience and sustainability targets ensures the longevity of such strategic planning beyond current political leadership. Each target is then quantified and measurable.

Summary

As risks and threats from climate change, pandemic, the global economy failures, and other threats are factored into a decision-making process, communities plan for long-term resilience and sustainability. The ability to communicate and implement a long-term vision for the organization is instrumental for an effective resilience and sustainability strategy. Moreover, resilience and sustainability planning assists in the integration of all the elements of the Quadruple Bottom Line, including the often-overlooked components of governance and social equity.

Local governments are confronted with many challenges, obstacles, and threats, including the

Table 2.1 Climate resilience report recommendations

Water	Energy	Build systems/ infrastructure	Transportation	Emergency preparedness
Climate resilience report recommendations				
Improve efficiency reduce water waste	Increase energy efficiency	Improve access to food sources	Change transportation culture around multimodal, mobility system for all residents.	Define hazard, emergency management, and disaster resilience
Use critical climate infrastructure	Reduce greenhouse gas emissions	Increase the number of certified sustainable buildings		
	Produce renewable energy			
Corresponding resilience and sustainability plan targets				
Reduce water consumption	Reduce City's consumption of gasoline, diesel, and natural gas	Increase number of community gardens	Increase miles of on-street bike lanes	All city employees involved in the National Incident Management System will maintain 100% of the training requirements to ensure preparedness.
Reduce stormwater discharge	Achieve 100% renewable energy in buildings	Improve access to farmer's markets	Develop miles of sidewalks	Regionalize emergency preparedness planning
Increase square footage of green roofs, pervious pavements, and parks	Reduce direct and indirect greenhouse gas emissions	Increase the number of sustainable redevelopment projects and buildings	Decrease total vehicle miles traveled by city employees	
Improve water reuse	Increase energy efficiency	Beach nourishment annually	Increase mix of alt. vehicle fuels	

effects of globalization, fiscal uncertainties, climate change, increased demands for services, and changing demographics. Contemporary organizations design their systems using effective sustainability and resilience planning to withstand external and internal pressures for maximum resilience in dynamic environments, including climate change, and growing economic, environmental, governance, and societal pressures (Alibašić 2018a, b, c, d, e, f). Furthermore, the local governments deploy sustainability as an organizational strategy (Alibašić 2018b). The geopolitical, financial, international, and socio-demographic trends, the evolving nature of technology, and a fundamental shift in values and unpredictable threats such as the recent COVID-19 pandemic generate challenges and opportunities for organizations.

A transformative, sustainable, and resilient organization focuses on continuous improvement while encouraging and enabling learning. As organizations evolve, they transform, continually reinvent themselves, invest in their development, and evaluate assets while identifying core competencies. In using the 5Is, The Initiate-Implement-Innovate-Inspect-Improve cycle in strategic resilience and sustainability planning, organizations facilitate a learning setting and continual development to grow and change and become more resilient in a dynamic environment. In identifying external pressures and charting a path forward, organizational leadership evaluates the necessity of having a strategic resilience and sustainability plan. Adopting a strategic resilience and sustainability plan and using it as a strategic plan and a guiding document increases synergies in implementing initiatives and executing the organizational and community-wide vision, mission, and goals.

Outcomes, Discussions, and Further Considerations

- Discuss the steps involved in setting the stage for organizational and citywide resilience and sustainability planning.
- Evaluate strategies that organizations have in place to address climate change and climate preparedness and readiness.
- Analyze the resources in place to implement resilience and sustainability planning.
- Assess the environmental protection programs, social justice, equity, economic opportunities, and good governance.
- Define the dissimilarity between strategic planning and resilience and sustainability planning.
- Analyze the benefits, opportunities, risks, and limitations of the resilience and sustainability planning process.
- Further align organizational resilience and sustainability strategies using the Quadruple Bottom Line (QBL) concepts and methods.

References

Alibašić H (2017) Measuring the sustainability impact in local governments using the quadruple bottom line. Int J Sustain Policy Pract 13(3):37–45

Alibašić H (2018a) Redesigning organizations for maximum resiliency in dynamic environments. In: Farazmand A (ed) Global encyclopedia of public administration, public policy, and governance. Springer International Publishing AG, Cham. https://doi.org/10.1007/978-3-319-31816-5

Alibašić H (2018b) Sustainability as organizational strategy. In: Farazmand A (ed) Global encyclopedia of public administration, public policy, and governance. Springer International Publishing AG, Cham. https://doi.org/10.1007/978-3-319-31816-5_3433-1

Alibašić H (2018c) Leadership strategies for embedding sustainability and resilience in organizations with an emphasis on sustainable energy. Creighton J Interdiscipl Lead 4(1):17–26

Alibašić H (2018d) Leading climate change at the local government level. In: Farazmand A (ed) Global encyclopedia of public administration, public policy, and governance. Springer International Publishing AG, Cham. https://doi.org/10.1007/978-3-319-31816-5_3428-1

Alibašić H (2018e) Ethics of resiliency in crisis management. In: Farazmand A (ed) Global encyclopedia of public administration, public policy, and governance. Springer International Publishing AG, Cham. https://doi.org/10.1007/978-3-319-31816-5

Alibašić H (2018f) Sustainability and resilience planning for local governments: the quadruple bottom line strategy. Springer, New York, NY. https://www.springer.com/us/book/9783319725673

Alibašić H (2020) Sustainable procurement. In: Farazmand A (ed) Global encyclopedia of public administration, public policy, and governance. Springer International AG, Cham. https://doi.org/10.1007/978-3-319-31816-5_3427-1

Alibašić H, Crawley W (2020) Applying good governance through policy labs: sustainable solutions for universities and local governments. Int J Soc Sustain Econ Soc Cult Cont 17(1):13–25. https://doi.org/10.18848/2325-1115/CGP/v17i01/13-25

Boswell MR, Greve AI, Seale TL (2012) Local climate action planning. Island Press, Washington, DC

City of New Orleans (2009) City of New Orleans carbon footprint report. https://www.nola.gov/getattachment/7aeaa7e4-7489-48f0-ac7b-4406c0926e79/Appendix-Ch-13-City-of-New-Orleans-Carbon-Footprin/

City of New Orleans (2015) Resilient New Orleans: strategic actions to shape our future city. http://resilientnola.org/wp-content/uploads/2015/08/Resilient_New_Orleans_Strategy.pdf

City of New Orleans (2017) Master plan. https://www.nola.gov/city-planning/master-plan/

City of New Orleans (2021) Resilient NOLA. http://resilientnola.org/

Cumo F, Garcia DA, Calcagnini L, Rosa F, Cumo F, Sferra AS (2012) Urban policies and sustainable energy management. Sustain Cities Soc 4:29–34. https://doi.org/10.1016/j.scs.2012.03.003

Droege P (2006) The renewable city: dawn of an urban revolution. Bull Sci Technol Soc 26:141–150. https://doi.org/10.1177/0270467606287531

Kennedy S, Sgouridis S (2011) Rigorous classification and carbon accounting principles for low and zero carbon cities. Energy Policy 39(9):5259–5268. https://doi.org/10.1016/j.enpol.2011.05.038

Lindfield M (2010) Cities: a global threat and a missed opportunity for climate change. Environ Urban ASIA 1(2):105–129. https://doi.org/10.1177/097542531000100202

Magis K (2010) Community resilience: an indicator of social sustainability. Soc Nat Resour 23(5):401–416. https://doi.org/10.1080/08941920903305674

Pierce JC, Budd WW, Lovrich NP Jr (2011) Resilience and sustainability in US urban areas. Environ Polit 20(4):566–584. https://doi.org/10.1080/09644016.2011.589580

United Nations Climate Change (UNCC) (2021a) What do adaptation to climate change and climate resilience mean? https://unfccc.int/topics/adaptation-and-resilience/the-big-picture/what-do-adaptation-to-climate-change-and-climate-resilience-mean

United Nations Climate Change (UNCC) (2021b) New elements and dimensions of adaptation under the Paris Agreement (Article 7). https://unfccc.int/topics/adaptation-and-resilience/the-big-picture/new-elements-and-dimensions-of-adaptation-under-the-paris-agreement-article-7

United Nations Framework Convention on Climate Change (UNFCCC) (2015) Paris agreement. https://unfccc.int/sites/default/files/english_paris_agreement.pdf

United States Census Bureau (2021) QuickFacts New Orleans, city. Louisiana. https://www.census.gov/quickfacts/fact/table/neworleanscitylouisiana/PST045219

Witcher BJ (2020) Absolute essentials of strategic management. Routledge, London

Identifying and Engaging the Internal and External Stakeholders, the Outcome and Target Champions and Collaborators

3

"When we deal with cities we are dealing with life at its most complex and intense."
Jacobs (1961, p. 372).

Overview

The third chapter includes an often-overlooked aspect of planning, stakeholder, partners and collaborators identification, and staff and community engagement. A sequence of the appropriate steps in identifying internal and external stakeholders for organizational and community strategic resilience and sustainability planning is described. The chapter offers in-depth insights into identifying community and organizational level engagement, internal and external stakeholders, organizers, partners, collaborators, and implementers throughout different stages of strategic planning. Like charting the planning process, the stakeholder engagement process is multi-layered, featuring diverse stakeholders, partners, and collaborators, both internal and external. Internal stakeholders include the elected and appointed officials, staff, and part-time employees. External stakeholders are residents, community members, partners, non-profit and business organizations and their representatives, educational institutions, boards and task forces, vendors, contractors, local, regional, state, and national organizations. The initial strategic resilience and sustainability planning action involves staff with defined roles and responsibilities and identifying target and outcome champions. The strategic resilience and sustainability planning process empowers employees to own the designated targets, promotes leadership and accountability, and furthers collaboration.

Keywords

Community engagement · Internal and external stakeholders · Outcome champions · Targets · Target champions · Change agents · Employee empowerment · Leadership · Integration · Embedding sustainability · Resilience · Embedment

Key Questions
The third chapter of the book answers the following underlying assumptions and questions:

- How to identify stakeholders, partners, collaborators, and outcome and target champions in the strategic resilience and sustainability planning?
- What is the relevance of a stakeholder evaluation and analysis and why is it valuable to organizations?

H. Alibašić, *Strategic Resilience and Sustainability Planning*, Sustainable Development Goals Series, https://doi.org/10.1007/978-3-030-91159-1_3

- What is the correlation between identifying the appropriate stakeholders, partners, collaborators, and champions and the acceptance of a strategic resilience and sustainability plan?
- Who are the critical stakeholders, collaborators, and partners in ensuring effective strategic resilience and sustainability plan implementation?
- How to engage a community and identify collaborators and partners for resilience and sustainability strategic planning?
- What are the state and national partnership opportunities for implementing resilience and sustainability projects?

Introduction

This third book chapter covers the stakeholder, collaborators, and partners identification, and the staff and community engagement, including designs necessary to implementing the strategies and initiatives. Stakeholder identification is often an overlooked aspect of strategic resilience and sustainability planning and includes identifying the internal and external stakeholders. Internal stakeholders include the elected and appointed officials, mayor, councilmembers, administrators, managers, staff, and part-time employees. External stakeholders, collaborators, and partners are residents, community members, partners, nonprofit, neighborhood and business organizations and their representatives, grassroots organizations, environmental and justice organizations, boards and task forces, vendors, contractors, and regional, state and national organizations.

A sequence of the appropriate steps in identifying internal and external stakeholders for organizational and community-wide resilience and sustainability strategic planning is explained. Similar to charting the initial resilience and sustainability strategic planning steps, the stakeholder engagement process is multilayered. It involves staff, managers, community leaders, collaborators, and partners, both internal and external. An overview of the synergies between departments, partners, community leaders, and staff working to achieve resiliency and sustainability-related objectives within an organization is explained. Definite subsections of stakeholder identification are described in-depth throughout the chapter. Freeman (1984, 1999) was credited with developing a stakeholder theory and later expanded the concept with the convergent stakeholder theory in an attempt to explain organizational dynamics.

Initially, the strategic planning process entails identifying appropriate staff, defining roles and responsibilities, and naming outcome and target champions. The collaborative planning process empowers employees to own the designated outcomes and targets, promotes leadership and accountability, and furthers the interaction between departments. Intentional strategic planning enables organizations to use a multifaceted, cross-sectoral approach to improve operational efficiency.

The primary drivers for prosperous resilience and sustainability programs are the ability of communities and organizations adapt to the changes in the governance, economic, environmental, and societal conditions. The process starts with vision and mission statements, an examination of the current conditions and available resources, existing resilience and sustainability programs and policies, and identifying stakeholders, both internally and externally. Some local government organizations have established resilience and sustainability programs, projects, and initiatives focusing on social equity, renewable energy, governance of resources, sustainable economic development, and environmental protection to name a few. With changes in elected leadership, there is a risk of termination of the resilience and sustainability initiatives.

Embedding resilience and sustainability strategies in an organization ensures continuity of the efforts regardless of the leadership changes. The strategic resilience and sustainability planning process is an instrument for multilevel stakeholder involvement to identify priorities for long-range goals and initiatives established. Moreover, it provides appointed administrators, managers, staff, and employees with a clear sense of ownership of the strategic goals and benchmarks for the expected results. The function of an organization's charge and consideration is to create a distinct focus on results, the ability to adjust elements as necessary, and a sense of ownership to attain long-term goals.

Through active stakeholder engagement, the organization's vision, mission, and objectives are communicated internally and externally. New programs put further pressures and add responsibilities to an existing workforce. Active stakeholder engagement and collaboration reduce ambiguities over the burden of adding new programs and projects to the current staff. The resilient and sustainable strategy ensures all staff, elected and appointed officials of local governments and external stakeholders, including community residents, businesses, contractors, and vendors, and partners are aware of the long-term strategic goals that an organization pursues. The organizations with the most momentum to advance strategic planning involve all staff, managers, and officials who are internal stakeholders and all external stakeholders.

Furthermore, having set resilience and sustainability steps in place, an organization demonstrates a commitment to reach those goals and establishes a vision for an improving and evolving organization. Once an organization's long-range goals, views, mission, and objectives are articulated through its resilience and sustainability strategic planning, the organization is bound to obtain and attain specific results. To determine community needs, and to identify and implement appropriate projects and programs, the planning process encompasses participation from across the spectrum of community sectors. In their research on regional governance and collaboration for resilience and sustainability outcomes Alibašić and Crawley (2018, 2020) showed the benefits of partnerships between the local governments and universities, and other participants. As each community faces its unique challenges and contains stakeholders, approaches need to be adjusted to fit those community features. Without recognizing the individual community needs and identifying the appropriate group of stakeholders, the resilience and sustainability strategies may falter and ultimately fail.

The novel nature of resilience and sustainability strategy accounts for external socioeconomic pressures and the present and future micro and macro-economic conditions. Likewise, sufficiently aligned strategies ensure the most crucial and relevant initiatives adhere to organizational vision while combining and adjusting for socioeconomic concerns. Resilient and sustainable strategy will ensure that internal and external stakeholders, partners, and collaborators are cognizant of the organization's future direction.

Climate Change and Pandemic Resilience

According to the National Centers for Environmental Information (NCEI), the US communities experienced weather and climate disasters with the total cost damages resulting from these 219 events exceeding $1.5 trillion, including hurricanes (NOAA 2018). These climate-related and recurring extreme weather events put an extraordinary strain on communities, and cities and counties are tasked with disaster response and post-disaster recovery.

The recent pandemic amplified difficulties of localized emergency management and disaster preparedness planning. Preliminary analysis of COVID-19 impact on local and state governments shows a significant impact on revenues and services (Auerbach et al. 2020; Sheiner and Campbell 2020). The resilience planning with adaptation, mitigation, and disaster preparedness elements are paramount for local government organizations as it prepares organizations for unpredictable shocks intensified by climate change and pandemics.

Resilience and Sustainability Planning and Change Agents

An effective strategic resilience and sustainability plan requires a change in perspectives and approaches to organizational development and improvement. When executing transformative changes with an ultimate goal of resilience and sustainability outcomes in organizations, the following represents indispensable elements in efficiently identifying internal stakeholders. The initial step involves the budget and finance employees, executive office administrators and managers, and departmental administrative staff evaluating strategies. Employees become change agents within an organization and advocate for resilience and sustainability strategic planning. Besides the top management, elected and appointed officials must fully support implementing changes and ensuring strategic resilience and sustainability plan accomplishments.

Implementing the change includes finding a starting point, attaching the benchmarks and measurements to existing efforts, constant communication, sharing progress information, involving staff and community, managing internal and external stakeholders, and embedding resilience and sustainability within the organization. Contemporary leaders, managers, and community members are dependent on fast-paced information that changes frequently and affects outcomes.

The leadership style and strategies deployed are necessary for organizations to survive and provide effective and efficient services. Effective leaders empower and enable employees to implement and efficiently execute strategies. Yukl et al. (2010) viewed the people as an integral component of organizational culture. Moreover, while Cambell et al. (2016) explained social assessment of urban parkland in resilience planning, Pfeffer (2010) focused on human aspects of sustainable organizations.

In local government organizations, municipal employees cooperate daily on a plethora of issues. Workers respond to concerned residents' complaints, political pressures, and delivery of service and represent the organization externally and internally. Employees represent the organization to the external customers. As resilience and sustainability measures in governance, economic, environmental, and social domains become part of the organizational dynamic, they facilitate teamwork, leadership, and transformational change. Mayrhofer et al. (2019) offered an insight into the reason why Human Resource Management must take context into consideration. In the context of strategic resilience and sustainability planning human resource management is a critical complementary component.

Empowering Employees to Champion Resilience and Sustainability

The strategic resilience and sustainability plan empowers employees to advance the identified targets and objectives, which leads to improved collaboration processes of governance in organizations. The targets and outcomes created with employee assistance and input are correlated to the budget outcomes. The outcome and target champions are designated to own specific outcomes and targets. Each champion of the resilience and sustainability outcome and target manages the process of accomplishing outcomes in partnership with counterparts in different departments or divisions to achieve outcomes set in a target.

The creation of the plan is a systemic process, not prescribed by the appointed and elected officials. The planning process is facilitated through the involvement of departments at various management levels of command, with delineation of duties between multiple departments to remove strict silos. Successful organizations encourage employees to seek more efficient ways to deliver a service and be more proactive in finding solutions. Murray and Holmes (2021) viewed employee empowerment as key to a sustainable workforce. Leaders empower employees by listening to their ideas and building on those. Pink (2010) observed effective leaders are good at listening and incorporating stories from their employees into their vision.

A capable resilience and sustainable leadership includes empowering employees to carry out the organization's vision. Such empowerment includes ethical behavior in the best interest of an organization where it becomes an ethical duty of employees to "challenge decisions that are misguided or unethical" (Yukl et al. 2010, p. 140) and becomes an administrative responsibility to plan for risks communities and organizations face, including climate change (Alibašić 2018, 2020) and pandemic (Bagwell 2020).

The strength of the organization is within and is carried by its employees. Cities and counties employees are encouraged to take more proactive, participatory, and vision-oriented roles to promote leadership and accountability. They are more likely to improve service delivery. In return, a personal leadership attitude enables employees to accomplish objectives with fewer resources, move forward on specific projects, and be open to alternative and innovative solutions, and use failure to improve service delivery.

In essence, resilience and sustainability strategic planning leads to self-leadership development through empowerment. As argued by Broman (2017, p. 3) "leadership development, in all domains, must integrate the application of systemic, systematic and strategic thinking and efforts." The strategic leadership application assists the organizations to be more competitive and to adapt to the changing and challenging environment (Ireland and Hitt 1999; Jefferies 2017; Hsieh et al. 2014).

Leadership in Contemporary Organizations

Modern organizations engage employees vertically and horizontally simultaneously to drive innovativeness through active employee engagement and equitable treatment. In organizations focused on resilience, sustainability, moderniza tion, and transformation, followers and leaders are necessary to successful outcomes. Often, there is a gap between the vision and financial reality, and that gap is filled with employees carrying the heavy burden of balancing the duties and be more proactive in their roles. Participative leadership theory offers an advantageous framework for active participant engagement in strategic resilience and sustainability planning. Busse and Regenberg (2018) analyzed the legacy of the authoritarian and participative leadership style and the effects on employee engagement and inclusiveness. Somech (2005) compared directive and participatory leadership styles. Somech's (2005) study did not find the two styles mutually exclusive, and Busse and Regenberg (2018) suggested the styles at the edges of the spectrum to be counterproductive.

Staff actively engaged in defining and crafting resilient and sustainable policies and implementing programs are better equipped to manage them. Developing leadership and maintaining a reliable and dedicated workforce is critical to strategic resilience and sustainability efforts. Organizational stakeholder engagement and identification of critical staff in delivering resilience and sustainability project are the most critical components of successful planning.

Identifying the Appropriate Teams for Climate Resilience Planning

Cities, towns, and communities along the coastal areas and inland in Florida, Louisiana, Texas, Alabama, and Mississippi face the more intense and frequent hurricanes and storms, with terrifying consequences. Identifying the appropriate climate resilience teams is required to disseminate climate change data and support adaptation strategies to strengthen community resilience. For instance, in the city of Biloxi in Mississippi, administrators identified the "city's emergency manager, floodplain manager, engineer, and Community Rating System coordinator" to disseminate climate change data and plan for climate resilience (City of Biloxi n.d.).

The city's impetus for recovery and resilience came to light in the aftermath of Hurricane Katrina. Simpson et al. (2010) discussed the damage from the hurricane on critical infrastructure on a community-wide scale in the cities of Biloxi and Gulfport, Mississippi. In analyzing Mobile Bay in Alabama, Bostick et al. (2017) argued for active stakeholders' engagement in resilience planning, stating it "utilizes the expertise of stakeholders and decision-makers to envision future conditions of importance" (p. 1182). Identifying the appropriate staff and stakeholders is a crucial step to the implementation of resilient and sustainable strategies.

In the city of Tallahassee (n.d.), Florida, city administration developed the Tallahassee Community Resilience Plan by engaging a broad spectrum of stakeholders and collaborators from the community, "interacting with residents and visitors at a variety of community events, ranging from neighborhood meetings to nonprofit roundtables," and "targeted meetings, workshops, and events" (para. 3). Furthermore, community outreach and engagement "included business and public agency leaders, college students and professors, youth at a summer camp, neighborhood associations, City employees, and many others," helping to "identify the different types of threats that the community is concerned about while also identifying approaches to move Tallahassee toward greater resiliency" (The City of Tallahassee n.d., para. 3).

The process of coastal resilience and sustainability strategic planning is critical in Florida due to ongoing climate change threats with serious consequences for the population (Alibašić and Morgan 2020). Cities must adjust their responses and harden the infrastructure to prepare for climate change impacts (Bloetscher et al. 2017). Besides, Nutt (2002) contended how implementation might be impeded with "unmanaged social and political concerns" (p. 98). It is imperative for strategic planning to include outcomes and targets to improve and afford training opportunities for employees to cope with an increased workload.

Organizations face transition and constant changes, and Uhl-Bien et al. (2007) speculated on adaptive leadership as "an emergent, interactive dynamic that produces adaptive outcomes in a social system" (p. 306). Transformation to resilience and sustainability is an adaptive process aligned with the pursuit of improved performance and outcomes. Tubbs and Schulz (2006) considered the organizational culture that embraces transformation, changes, and "continuous learning, building mechanisms to create and sustain change efforts, managing the change process, developing change agents, and encouraging individuals as well as a structural change in the organization" (p. 32). These factors are essential for nurturing leaders and empowering employees, where employees are decisive stakeholders in implementing policies and programs.

Organizations empower employees to advance the identified targets and objectives in a strategic resilience and sustainability plan, which leads to improved collaboration processes of governance in organizations. The targets created with the assistance and input from employees are correlated to the budget outcomes. Outcome and target champions are designated to own and be responsible for specific outcome and targets. Each champion of the resilience and sustainability target manages the process of accomplishing outcomes in partnership with counterparts in different departments or divisions.

The creation of the plan is a systemic process, not prescribed by the appointed and elected officials. The planning process is facilitated by the involvement of departments at various management levels of command, with delineation of duties between multiple departments to remove strict silos. Robust organizations encourage employees to seek more efficient ways to deliver a service and be more proactive in finding solutions. Murray and Holmes (2021) viewed employee empowerment as key to a sustainable workforce. Leaders empower employees by listening to their ideas and incorporating suggestions from staff in the planning processes. Pink (2010) observed effective leaders are good at listening and incorporating stories from their employees into their vision.

Empowering employees to own and execute specific projects in ther work portfolio is a critical objective in a strategic resilience and sustainability plan. Empowerment includes ownership of the vision, goals, and mission the organization sets in place. Administrators and managers have a responsibility and duty to plan for risks and threats, including climate change (Alibašić 2018, 2020) and pandemics (Bagwell 2020).

The organizations strength is internal and is carried by its employees. Empowered workers and staff take proactive, participatory and vision-oriented roles, advance leadership, embrace accountability, and improve service delivery. In return, leadership enables employees to accomplish objectives with fewer resources, advance specific projects, and be open to alternative and innovative solutions, and use failure to improve service delivery.

In essence, resilience and sustainability strategic planning leads to self-leadership development through empowerment. As argued by Broman (2017, p. 3), "leadership development, in all domains must integrate the application of systemic, systematic and strategic thinking and efforts." Strategic leadership and management assist the organizations to be more competitive and to adapt to the changing and challenging environment (Ireland and Hitt 1999; Jefferies 2017; Hsieh et al. 2014).

Nurturing future leaders within organizations requires a strong commitment to diversity and diverse cultures. Often, the challenging economic times prompt municipal governments toward adaptive leadership, fostering resilience, sustainability, adaptability, and positive transformation. Leadership from within is nurtured by current elected and appointed officials to prepare future leaders for succession.

Assigning the Outcome and Target Champions

The strategic resilience and sustainability plan sets specific objectives using measurable targets and timelines to accomplish initiatives. The strategic plan incorporates the assignment of responsibilities to departmental staff. For illustration, under the environmental outcomes and goals, one of the targets could be to increase the number of trees planted on private property by 40,000 by 2030, using 2022 as a benchmark year of 5,000 trees a year. While the first champion assigned to this target is an employee from the division of the public works department, staff from other divisions such as parks and recreation, forestry, and streets are instrumental in supporting the outcomes related to this particular target. The synergy leads to improved collaboration, leveraging of resources, and ultimately expediting progress.

Administrators encourage and nurture leadership and ownership. However, a similar strategy is applied at a community level to increase the chances of successful integration and implementation of a strategic plan. A proactive process of community engagement, surveys, and other tools for seeking and including community input is deployed. The targets generated are attached to the budget outcomes and are championed and maintained by employees. Each champion of the resilience and sustainability target ensures the outcome delivery.

Organizations transitioning to resilience and sustainability strategic planning engage the employees through a planning process. Covey (2009) noted the difficulty of accepting the plans imposed on employees. Strategies for effective employee empowerment and engagement include controlling outcomes, engaging in planning processes, updating progress, and making recommendations for improvements. Leaders recognize the relevance of staff and employees for implementation and transitions to improved operations. Organizations advancing and empowering employees include coaching and offer training to staff and workers to ensure professional development and growth.

Active participation and leadership development remove underlying tensions and allow for positive feedback between organizational levels hierarchy. As pointed out by Friedman (2004), "people tend to misperceive dynamic feedback in an organization" (p. 111). In assigning the champions for each target in the plan, organizations ensure staff accountability in attaining their outcomes. Linkov et al. (2013) discussed a resilience measurement for an actionable policy framework. Resilience and sustainability measurements and initiatives are connected to specific targets and outcomes. To create a resilient and sustainable community, every staff is accountable for decisions and actions impacting others. The outcome and target champions measure, observe the assigned targets' progress, and attain targets and outcomes.

Community Engagement

Community engagement is an indispensable facet of resilience and sustainability strategy for local governments. The informed community appreciates how resilience and sustainability planning improves service delivery and how it advances organizational efficiency. The engaged community members are more likely to be supportive of resilience and sustainability actions. The solution to community engagement is to identify the organization's uniqueness and the community and apply strategies to fit specific characteristics.

Community's participation, partnerships, stakeholder engagement, and the commitment to the pursuit of resilience and sustainability are essential to a prosperous plan implementation and completion. As cities transition their plans to account for climate change, pandemic risks, and other uncertainties and risks, they seek input and support from the community using an innovative process of engaging residents and business owners through transformation advisory groups, and resilience and sustainability task forces. Mascarenhas et al. (2015) argued for the inclusion of a broad range of perspectives to be represented by diverse stakeholders in the planning process.

The depth and breadth of identified stakeholders increase the likelihood of robust implementation of resilience- and sustainability-related activities. However, progress requires active integration throughout the entirety of the planning process beyond the mere inclusion of community members in the early planning stages.

The planning outcome achievements in the community are dependent on the collaborative efforts of all department in all aspects of services delivery, including but not limited to energy management, waste minimization, economic growth, human resource development, diversity, community education, public health, environmental protection, with support from the non-profit, residential, academic, and business community representatives.

Hitchcock and Willard (2008) proposed "sustainability solutions often require an interdisciplinary, multi-stakeholder approach, involving people from across the organization or even from multiple organizations" (p. 119). Similarly, Bostick et al. (2017) saw advantages in the engagement of diverse stakeholders in coastal resilience planning facilitated through dialogues.

Early strategic resilience and sustainability planning process requires building collaborative relationships to complement city administrators, staff, managers, directors, and elected and appointed officials. Resilience and sustainability action task force, workgroups, action teams, and cooperative groups are advantageous in achieving the actionable objectives. Cities have been effectively utilizing strategies for engaging community members to advance resilience and sustainability initiatives.

For example, the city of Beaverton, OR, hosted a series of community engagement forums to discuss sustainability efforts, including the roundtable meetings, considering the partnership opportunities with the regional, state, and federal agencies and stakeholders to create a sustainable community (NLC 2013). The city's concerted effort included roundtables for a meaningful and organic community engagement, seeking input, and providing opportunities for multiple stakeholders to contribute toward the resilience and sustainability of the community. The community engagement continued throughout the years resulting in the annual vision plan (City of Beaverton 2017).

Additionally, Sarasota and Manatee Counties consider civic engagement as one of their top priorities included in the resilience dashboard (Sarasota 2021; Transition Sarasota 2021). The greater the participation in planning processes by businesses, residents, and other interest groups, the more integrated and impactful the strategic resilience and sustainability plan is for communities. Resilience and sustainability strategic planning is viewed through the lens of civic engagement of interest groups pursuing resilient and sustainable outcomes. Resilient Boston strategy relies heavily on participatory engagement offering "residents a meaningful role in decision-making processes," and facilitating "cross-departmental partnership" (City of Boston 2017, p. 62). Resilient Boston strategy was developed with the community and stakeholder input (City of Boston 2016). Under vision 2 of the strategy, the City of Boston (2017, p. 63) envisions collaborative, proactive governance, prioritizing the "community-led processes and community partnerships, where City services are delivered equitably to people and communities, and the City government reflects the diverse culture and people it serves," with three distinct goals:

- Ensuring employment equity and better serving all Boston residents by increasing the representation of the city's diverse population in the City government.
- Enhancing decision-making capacity in City government by bringing together residents' and government representatives' knowledge and skills to better develop policies, practices, and processes.
- Improving the collaboration, evaluation and delivery of City services to better meet the needs of all Boston residents.

Community engagement strategies may take months and, in some cases, years. For example, the City of Kalamazoo (2021) in Michigan started its draft of the sustainability action plan in July of 2019 with the project kick-off and the community-wide survey, followed by sustainability advisory and neighborhood meetings in October and December of 2019, and then in 2020 with survey analysis and focus groups, department meetings and information gathering, data collection, data analysis, data mapping, gap inventory, and greenhouse gas emissions inventory. In 2021, the city continued with department meetings and drafting the inventory, action table, and priorities to complete a draft sustainability action plan in August of 2021, followed by additional public engagement and approvals (City of Kalamazoo 2021). Communities with resilient and sustainable initiatives have a participatory citizenry and seek to achieve the social, economic, environmental, and governance outcomes through citizen engagement and input.

Resilient Communities and Organizations Spotlight: Community Engagement for Resilient DC

The Washington DC's Office of Resilience was established to deal with "catastrophic shocks and chronic stresses to ensure that DC thrives in the face of change," and to endure human-made and natural calamities and confront "the social challenges that come with being a fast-growing city" (Office of Resilience 2017). Some of the significant stressors and shocks identified by the DC Office of Resilience are threats of potential terrorist attacks, road congestions, and climate change-induced extreme weather, including heatwaves.

The DC Office of Resilience utilized an extensive community engagement process, including surveys to identify strengths and weaknesses in the community. Its Resilience Strategy is divided into four major goals: Inclusive Growth, Climate Action, Smarter DC, and Safe and healthy Washingtonians (Government of The District of Columbia 2019). The strategy reaffirms the District's commitment to Paris Climate Accord and to lead in combating climate change envisioning a prepared and ready DC "for the impacts of climate change and where residents and businesses take bold action to combat its causes" (Government of The District of Columbia 2019, p. 5). Brugmann (2012) discoursed scaling climate adaptation strategies in urban settings, describing resilience in the function of expectable performance by an "urban asset, location and/or system" (p. 217).

Cities provide services and identify policy options in the organization's best interest and community at large. Seeking resident and business input through active civic engagement is vital for a resilient and sustainable community. The Office of Resilience in DC's ultimate goal is developing a "resilience strategy [that] will be a holistic, action-oriented plan to build partnerships and alliances as well as financing mechanisms, and will pay particular attention to meeting the needs of vulnerable populations" (Office of Resilience 2017).

Resilient Communities and Organizations Spotlight: Boston Governance and Public Engagement

City of Boston (2017) adopted the Resilient Boston report with a subtitle An Equitable and Connected City. The report is divided into four sections, and the second section focuses on collaborative, proactive governance with five initiatives: to improve equity in city employment, drive innovation in community engagement, pioneer interactive resilience platform, create a funding pipeline for resilience initiatives in the community, and leverage city data to advance equity. The city of Boston prioritized community engagement, diversity, transparency, public input, partnership, equity, and collaborations, and other aspects of good governance and accountability.

In identifying internal stakeholders and implementers, collaborators, partners, and external supporters, local governments chart the resilience and sustainability strategic plan process with an understanding of its community-wide implications. As part of the process of community engagement, municipalities seek partnership opportunities for continuity and adaptability in planning. While the local governments' resilience and sustainability planning relates to organizational operations and resources, it also concentrates on the broader implication of resiliency and sustainability in the region and impact on the community.

Collaborative resilience and sustainability planning is accomplished through a variety of measures and initiatives, including energy audits, renewable energy projects, micro-loans initiatives, affordable housing opportunities, greenhouse gas emissions, life cycle projects, waste minimization, climate change actions, pandemic responses, efficiency improvements in residential homes, and promotion of recycling through local businesses and chambers of commerce.

Resilient Communities and Organizations Spotlight: Cincinnati, Cleveland, and Columbus, Ohio

In 2013, the local government in Cincinnati developed the Green Cincinnati Plan, focusing on sustainability. Later, the city expanded the plan to emphasize equity (City of Cincinnati 2018). A diverse, collaborative effort aims to promote and share the best sustainability and equity practices in planning and operations. Through a series of four public meetings, the Office of Environmental Quality in Cincinnati engaged more than 200 members of the Green Umbrella Action Teams and other community members to assist and provide input and recommendations for the Green Cincinnati Plan.

The Green Umbrella is a regional sustainability alliance in the Cincinnati region. A steering committee consisting of representatives from both the public and private sectors provided leadership in reviewing and finalizing the recommendations from the Green Umbrella Action Teams. The steering committee made the final recommendations to the city administrators after holding a charrette (City of Cincinnati 2013; City of Cincinnati 2018). Local education institutions, private sector companies, and the city seek the best solutions to region-wide issues, each working towards a resilient community and developing a sustainability plan.

The city of Cleveland leaders emphasized the importance of resilient neighborhoods. Through its funding intermediary, the Cleveland Neighborhood Progress, the city emphasized collaboration to accomplish more resilient and sustainable neighborhoods. The city relied on broad-based collaboration, partnerships, local participation, inclusiveness, and empowering residents for positive neighborhood outcomes (Cleveland Neighborhood Progress 2021a). One of its major goals is climate resiliency and sustainability in neighborhoods with an emphasis on sustainability and mitigation strategies, social cohesion, inclusiveness, access to green spaces and parks, and resilient land use and reuse (Cleveland Neighborhood Progress 2021b; City of Cleveland 2019).

Finally, the city of Columbus in Ohio promotes the community resilience coalition. Through the Columbus Community, Action, Resilience and Empowerment (CARE) Coalition, the city intends to assist individuals, families, and neighborhoods in times of crisis and provides a list of services, including information about the neighborhoods and county-wide community resources for COVID-19 (City of Columbus 2021).

Regional, National, and International Partnership Opportunities

Collaborating with a regional or a state or a national organization is a beneficial form of partnership for delivering and implementing resilience- and sustainability-related initiatives. There are numerous local, regional, state, national, and international organizations involved in resilience and sustainability-related activities such as the Florida League of Cities (FLC), ICLEI – Local Governments for Sustainability, US Conference of Mayors (USCM), National League of Cities (NLC), and other organizations.

Cities may partner to share best practices through regional partnership opportunities. For example, 27 member counties and cities, and nine agencies are members of the East Central Florida Regional Planning Council committed to regionalism and regional resilience collaboration efforts. Member cities and counties are Cape Canaveral, Cocoa, Cocoa Beach, DeLand, Deltona, Indian Harbour Beach, Kissimmee, Lake Helen, Mascotte, Melbourne Beach, Mount Dora, New Smyrna Beach, Oak Hill, Orange City, Orlando, Ormond Beach, Palm Bay, Ponce Inlet, Rockledge, Satellite Beach, Titusville, Winter Park, Lake County, Orange County, Osceola County, Volusia County, and Brevard County (East Central Florida Regional Planning 2021).

The East Central Florida Regional Planning (2021) identified the following three pillars of resilience "(people) Health + Equity, (places) Build Infrastructure + Natural Environment, and (prosperity) economic resilience" (para. 2). The main goals are "reducing the carbon footprint, risks and vulnerabilities utilizing emergency management, and increasing efforts toward sustainability, region-wide" (East Central Florida Regional Planning 2021, para. 2). Furthermore, regional planning efforts aim to reduce unnecessary competition between local governments and support regional economic development and growth.

In the north of the United States and Canada, bordering Great Lakes, over a hundred cities, members of the Great Lakes-Saint Lawrence Cities Initiative (GLSLCI), promote a multistate, bi-national strategy to curb environmental and climate change threats to the largest body of fresh water on Earth, and Asian carp and other invasive species threatening the Great Lakes ecosystem (GLSLCI 2021a, b). Beyond water quality advocacy in addressing nutrients and algae bloom, biodiversity, micro-plastics, oil transport, and nuclear energy, the organization undertook a myriad of other initiatives and actions to support the members' programs for climate adaptation and mitigation, and reduction of microplastic in the Great Lakes and Saint Lawrence water basin (GLSLCI 2021a).

In DC, the GLSLCI members advocate for the Great Lakes Compact, seeking funds and plans for the Asian carp threats to the Great Lakes and advocating for separation of the Great Lakes and Mississippi River watersheds to safeguard from aquatic invasive species. Collaborative resilience and sustainability planning initiatives in the region, energy audits, energy efficiency improvements in neighborhood homes, increased recycling through local economic development incentives, and the regional climate action reporting further the relevance of partnerships to achieve prosperous sustainability and resilience-related outcomes. By using the system-wide integrated approach to resilience and sustainability issues, the GLSLCI member mayors and staff from the United States and Canadian cities form an alliance to preserve and protect the precious Great Lakes and Saint Lawrence water assets and communities.

Summary

The formalized process of strategic resilience and sustainability planning is critical. Triumphant integration and implementation of resilience and sustainability initiatives and practices require assessment and sensitivity related to an organization's culture, the professionals within that organization, and the stakeholders' demographics. The universal language of resilience and sustainability strategic planning conveys the well-known principles that resonate with the implementation of the plan.

Moreover, resilience efforts are flexible with strong leadership commitment. Finally, the strategic resilience and sustainability planning leads to improved outcomes. The renewed planning process encompasses governance, economic, environmental, and social objectives to improve operational efficiency and community resilience and creates a framework for meeting goals and objectives. By expertly identifying and engaging key stakeholders, collaborators, and partners cities, counties and other organizations, seek acceptance and support for essential community initiatives. Moreover, organizational strategic resilience and sustainability planning includes defining the strategies for initiatives, expressing vision, goals, and objectives, communicating them to its internal and external stakeholders and collaborators, and identifying target and outcome champions.

The sense of ownership over an organization's objectives affords internal and external stakeholders to seek the most effective outcomes for initiatives. Additionally, a strategic resilience and sustainability plan expresses a vision while empowering stakeholders, namely employees, to exceed expectations in implementing initiatives. It takes into consideration possible partnerships and invites future collaborative opportunities. As noted in several examples of initiatives undertaken by communities and organizations, strategic sustainability and resilience plans establish the course for an organizational and community future.

Outcomes, Discussions, and Further Considerations

- Analyze the significance of gaining internal and external stakeholders' support in resilience and sustainability strategic planning.
- Evaluate and analyze stakeholders, partners, and collaborators for resilience and sustainability planning.
- Discuss the organizational leadership pitfalls in stakeholders' and community engagement.
- Create a stakeholder assessment of external and internal factors in the resilience and sustainability planning process.
- Assess the connection of stakeholder analysis with that of socioeconomic, environmental, and resources examination.
- Understand the importance of selecting and appointing target and outcome champions, and collaborators and partners.

References

Alibašić H (2018) Sustainability and resilience planning for local governments: the quadruple bottom line strategy. Springer, New York, NY. https://www.springer.com/us/book/9783319725673

Alibašić H (2020) The administrative and ethical considerations of climate resilience: the politics and consequences of climate change. Public Integr. https://doi.org/10.1080/10999922.2020.1838142

Alibašić H, Crawley W (2018) Developing and continuing sustainability-related academic programming: observations of emerging practices. Creighton J Interdiscipl Lead 4(1):27–34

Alibašić H, Crawley W (2020) Applying good governance through policy labs: sustainable solutions for universities and local governments. Int J Soc Sustain Econ Soc Cult Cont 17(1):13–25. https://doi.org/10.18848/2325-1115/CGP/v17i01/13-25

Alibašić H, Morgan JD (2020) Coastal climate readiness and preparedness: comparative review of the state of Florida and Cuba. In: Perception, design and ecology of the built environment: a focus on the global south. Springer, New York, NY. https://www.springer.com/gp/book/9783030258788

Auerbach A, Gale B, Lutz B, Sheiner L (2020) Fiscal effects of COVID-19. https://www.brookings.edu/wp-

content/uploads/2020/09/Auerbach-et-al-conference-draft.pdf
Bagwell J (2020) Leading through a pandemic: adaptive leadership and purposeful action. J School Admin Res Dev 5(S1):30–34. ISSN: 2470-8496 Print/ISSN: 2470-850X Online ojed.org/jsard
Bloetscher F, Hoermann S, Berry L (2017) Adaptation of Florida's urban infrastructure to climate change. In: Chassignet EP, Jones JW, Misra V, Obeysekera J (eds) Florida's climate: changes, variations, & impacts. Florida Climate Institute, Gainesville FL, pp 311–338. https://doi.org/10.17125/fci2017.ch11
Bostick TP, Holzer TT, Sarkani S (2017) Enabling stakeholder involvement in coastal disaster resilience planning. Risk Anal 37(6):1181. https://doi.org/10.1111/risa.12737
Broman G, Robert K, Collins TJ, Basile G, Baumgartner RJ, Larsson T, Huisingh D (2017) Science in support of systematic leadership towards sustainability. J Clean Prod 140:1–9
Brugmann J (2012) Financing the resilient city. Environ Urban 24:215–232. https://doi.org/10.1177/0956247812437130
Busse R, Regenberg S (2018) Revisiting the "authoritarian versus participative" leadership style legacy: a new model of the impact of leadership inclusiveness on employee engagement. J Leadersh Org Stud 26(4):510. https://doi.org/10.1177/2F1548051818810135
Cambell LK, Svendsen ES, Sonti NF, Johnson ML (2016) A social assessment of urban parkland: analyzing park use and meaning to inform management and resilience planning. Environ Sci Pol 62:34–44. https://doi.org/10.1016/j.envsci.2016.01.014
City of Beaverton (2017) Vision plan FY-2016-2017. https://www.beavertonoregon.gov/DocumentCenter/View/16061/Community-Vision?bidId=
City of Biloxi (n.d.). City of Biloxi, Mississippi, Program For Public Information. Retrieved from: https://masglp.olemiss.edu/projects/crsppi/files/biloxi-ppi.pdf
City of Boston (2016) Boston's preliminary resilience assessment: the beginning of a shared journey to healing, connections, and actions. https://www.boston.gov/sites/default/files/embed/p/prabrochure-update_printcopy.pdf
City of Boston (2017) Resilient Boston: an equitable and connected city. https://www.boston.gov/sites/default/files/file/document_files/2017/07/resilient_boston.pdf
City of Cincinnati (2013) Green Cincinnati plan. https://www.cincinnati-oh.gov/oes/linkservid/6CE53223-9206-9F36-DB7FA3444F16A1A0/showMeta/0/
City of Cincinnati (2018) 2018 Green Cincinnati plan. https://www.cincinnati-oh.gov/sites/oes/assets/File/2018%20Green%20Cincinnati%20Plan(1).pdf
City of Cleveland (2019) Cleveland climate action plan. Budling thriving and healthy neighborhoods. https://d3n8a8pro7vhmx.cloudfront.net/sustainablecleveland/pages/149/attachments/original/1461798511/Cleveland_Climate_Action_Plan.pdf?1461798511
City of Columbus (2021) Columbus CARE coalition. https://www.columbus.gov/publichealth/programs/Neighborhood-Services/Community-Resilience-Coalition/
City of Kalamazoo (2021). Imagine Kalamazoo 2025: Kalamazoo sustainability strategy. Retrieved from http://www.imaginekalamazoo.com/projects/sustainability/
City of Tallahassee (n.d.) Community resilience plan. https://www.talgov.com/publicsafety/resplan.aspx
Cleveland Neighborhood Progress (2021a). http://www.clevelandnp.org/strategicplan/
Cleveland Neighborhood Progress (2021b) Climate resiliency - resilient Cleveland. http://www.clevelandnp.org/resilientcleveland/
Covey SMR (2009) How the best leaders build trust. Leadership now. http://www.leadershipnow.com/CoveyOnTrust.html
East Central Florida Regional Planning (2021) Programs. https://www.ecfrpc.org/resiliencecollaborative
Freeman RE (1984) Strategic management: a stakeholder approach. Pitman Publishing, Boston, MA
Freeman RE (1999) Divergent stakeholder theory. Acad Manag Rev 24(2):233. https://doi.org/10.5465/amr.1999.1893932
Friedman S (2004) Learning to make more effective decisions: changing beliefs as a prelude to action. Learn Organ 11(2):110–128
Government of The District of Columbia (2019) Resilient DC. A strategy to thrive in the face of change. https://app.box.com/s/d40hk5ltvcn9fqas1viaje0xbnbsfwga
Great Lakes Saint Lawrence Cities Initiative (GLSLCI) (2021a) Initiatives https://glslcities.org/initiatives/
Great Lakes Saint Lawrence Cities Initiative (GLSLCI) (2021b) Members. https://glslcities.org/member-cities/
Hitchcock DE, Willard ML (2008) The step-by-step guide to sustainability planning: how to create and implement sustainability plans in any business or organization. Earthscan, London
Hsieh MH, Huang HY, Seah M (2014) Leader driven organizational adaptation. Manag Decis 52(8):1410–1432. https://doi.org/10.1108/MD-07-2013-0380
Ireland D, Hitt M (1999) Achieving and maintaining strategic competitiveness in the 21st century: the role of strategic leadership. Acad Manag Exec 13(1):43–57. Business Source Premier Database
Jacobs J (1961) The death and life of great American cities. Vintage Books, New York, NY
Jefferies SS (2017) Adaptive leadership in a socially revolving world: a symbolic interactionist lens of adaptive leadership theory. Perform Improv 56(9):46–50. https://doi.org/10.1002/pfi.21741
Linkov I, Eisenberg DA, Bates ME, Chang D, Convertino M, Allen JH, Flynn SE, Seager TP (2013) Measurable resilience for actionable policy. Environ Sci Technol 47(18):10108–101010
Mascarenhas A, Nunes LM, Ramos TB (2015) Selection of sustainability indicators for planning: combining stakeholders' participation and data reduction techniques. J Clean Prod 92:295–307

Mayrhofer W, Gooderham PN, Brewster C (2019) Context and HRM: theory, evidence, and proposals. Int Stud Manag Organ 49(4):355–371. https://doi.org/10.1080/00208825.2019.1646486

Murray WC, Holmes MR (2021) Impacts of employee empowerment and organizational commitment on workforce sustainability. Sustainability 13:3163. https://doi.org/10.3390/su13063163

National League of Cities (NLC) (2013) Revitalization roundtable kicks off local- federal conversations on sustainable manufacturing, job creation and redevelopment in Beaverton, Oregon. http://www.nlc.org/article/revitalization-roundtable-kicksoff-local-federal-conversations-on-sustainable

National Oceanic and Atmospheric Administration (NOAA). National Centers for Environmental Information (2018) Billion-dollar weather and climate disasters: overview. https://www.ncdc.noaa.gov/billions/

Nutt PC (2002) Why decisions fail: avoiding the blunders and traps that lead to debacles. Berrett-Koehler Publishers Inc, San Francisco, CA

Office of Resilience (2017) Resilient Washington DC: about resilient DC. https://resilient.dc.gov/page/about-resilient-dc

Pfeffer J (2010) Building sustainable organizations: the human factor. Acad Manag Perspect 24(1):34–45. 12 p; Business Source Premier Database

Pink D (2010) Analytics are out, autonomy is in. Walden Alumni Mag 5(2):27–28. Summer/Fall 2010. Walden University. https://www.waldenu.edu/-/media/Walden/files/newsroom/alumni-magazines/waldenmagazine-5-2.pdf

Sarasota County (2021) Resilient American communities. Sarasota and Manatee Counties. https://resiliencesystem.org/dashboards/sarasota-county/

Sheiner L, Campbell S (2020) How much is COVID-19 hurting state and local revenues? Brookings. https://www.brookings.edu/blog/up-front/2020/09/24/how-much-is-covid-19-hurting-state-and-local-revenues/

Simpson DM, Lasley CB, Rockaway TD, Weigel TA (2010) Understanding critical infrastructure failure: examining the experience of Biloxi and Gulfport, Mississippi after Hurricane Katrina. Int J Crit Infrastruct 6(3):246–276

Somech A (2005) Directive versus participative leadership: two complementary approaches to managing school effectiveness. Educ Adm Q 41:777. https://doi.org/10.1177/0013161X05279448

Transition Sarasota (2021) Transition Sarasota. https://www.transitionsrq.org/our-story

Tubbs S, Schulz E (2006) Exploring a taxonomy of global leadership competencies and metacompetancies. J Am Acad Bus Camb 8(2):29–34

Uhl-Bien M, Marion R, McKelvey B (2007) Complexity leadership theory: shifting leadership from the Industrial Age to the knowledge era. Leadersh Q 18(4):298–318

Yukl G, George JM, Jones GR (2010) Leadership: building sustainable organizations (Laureate Education, Inc., custom ed.). Custom Publishing, New York, NY

4 Measuring, Tracking, Observing, Scrutinizing and Reporting the Resilience and Sustainability Outcomes and Results

"Proper internal control: The third and last necessity to achieve success in business is to have a system of internal control. With a system of internal control, business transactions are recorded in such a systematic way that one may understand each one of them at glance."

Fra Luca Paciolli (1494, pp.1–2) Particularis de Computis et Scripturis

Overview

Chapter four defines the exact measurements, tracking tactics, observing, scrutinizing, and reporting methods using the quadruple bottom line strategy. It develops a progress report to ensure accountability, answerability, transparency, and good governance. The foundation for strategic resilience and sustainability planning is the ability of staff in organizations to measure, track, monitor, and communicate resilience and sustainability-related outcomes. In guaranteeing the auspicious implementation of strategic resilience and sustainability plan, organizations analyze, evaluate, and disseminate information about the outcomes and goals and track and measure them over a period of time. The most appropriate and effective tracking, monitoring, scrutinizing, and reporting the initiatives and activities include evaluating and assessing the goals, outcomes, objectives, and targets.

Keywords

Measuring · Observing · Tracking · Reporting · Targets · Benchmark data · Results · Target champions · Budget alignment · Qualitative data · Quantitative data · Reporting · Disseminating and communicating outcomes · Progress update · Baseline data · Good governance · Transparency · Accountability Answerability

Key Questions

This fourth chapter of the book answers the following assumptions and questions:

- Why do organizations measure resilience and sustainability outcomes and report the results?
- What do organizations measure to increase resilience and become sustainable?
- What are the objective and relevant resilience and sustainability measurements?
- What targets are relevant to organizational resilience and sustainability strategic planning?
- What are the differences between the indicators and targets?
- What are the connections between measuring, tracking, observing, scrutinizing and reporting the resilience and sustainability activities in operations in diverse communities?

H. Alibašić, *Strategic Resilience and Sustainability Planning*, Sustainable Development Goals Series, https://doi.org/10.1007/978-3-030-91159-1_4

- What type of reports are the most effective to communicate the progress of resilience and sustainability-related initiatives?
- How do organizations ensure transparency, good governance, answerability, and accountability through resilience and sustainability progress reports?

Introduction

The fourth chapter of the book examines the functions and consequences of the measurements of resilience and sustainability activities, projects, and initiatives. The foundation for strategic resilience and sustainability planning is the ability of staff in organizations to measure, track, observe, scrutinize, and report the resilience and sustainability-related outcomes. This chapter covers measuring, tracking, observing, and reporting process using the Quadruple Bottom Line strategy and development of a progress report to ensure accountability, transparency, answerability, and good governance.

In guaranteeing the prolific implementation of resilience and sustainability initiatives, organizations analyze, evaluate, and disseminate information about the outcomes and objectives and track and measure them constantly. The assessment of appropriate and effective ways of tracking, observing, and reporting the initiatives includes evaluating the resilience and sustainability goals, objectives, outcomes, and targets.

A proactive approach to measuring and observing outcomes and targets in the strategic resilience and sustainability plan ensures the departments and city staff are held accountable through ownership of the outcomes and targets.

Progress reports available online or in print guarantee an additional answerability for the resilience and sustainability-related outcomes. In balancing sustainability and resilience objectives, organizations report progress on each target annually or bi-annually. Measuring outcomes in each initiative and activity are paramount to implementing the objectives of the plan. The design of resilient and sustainable programs does not warrant more efficacy unless the programs and initiatives are effectively tracked, measured, observed, scrutinized, and reported.

Quantifying and reporting results are critical to guaranteeing answerability, accountability, and transparency. An example of effective measurements of resilience-related outcomes is the local governments' institutionalization of greenhouse gas (GHG) emissions reduction, often referred to as a carbon footprint reduction. Cities, counties, townships, and other organizations express the commitments for carbon reduction through targets in a strategic resilience and sustainability plan, observation of related results, and publication of greenhouse gas emission reduction activities.

Administrators measure, track, observe, and report outcomes related to GHG emission reduction for the entire organization and community. Benefits from a reduced carbon footprint are interpreted as benefits for the community, including direct savings or cost avoidance in energy consumption. A record of measurable outcomes contributing to a reduction of GHG emissions includes waste minimization, energy efficiency improvements, renewable energy production, installation of electric vehicle charging stations, electric vehicles purchase programs, increased recycling availability, low-impact development, better stormwater management, the addition of bike lanes, and aggressive tree-planting programs.

Organizations' analysts convert the data and information about the ecological advantage from carbon footprint reduction in operations into equivalent gains for the entire community. Organizations committed financially and otherwise to the resilience goals, and practical application and implementation of those efforts feature their progress prominently toward ultimate positive outcomes. Local governments connect resilience and sustainability planning directly to the budget process and fiscal plans.

Staff and administrators track, measure, observe, scrutinize, and report results. By measuring, tracking, scrutinizing, and reporting data, administrators identify opportunities for cost-saving measures. For example, cities measure and report waste minimization and energy efficiency initiatives, using a secondary layer of reporting the community's emissions from residential, commercial, industrial, and transportation-related activities. Through a detailed measurement of available data, administrators comprehensively explain the greenhouse gas emissions in cities or counties' facilities and fleets, using the CO_2 equivalent.

Finally, organizations indicate and substantiate commitment to resilience and sustainability planning to connect resilience and sustainability planning directly to the budget process and fiscal plans. The resources and templates to feature and showcase progress are available to local governments seeking them.

Resilient Communities and Organizations Spotlight: Resilient and Resource-Efficient Alachua County, FL

One of the early adopters of the term resilience in planning, in 2000 the Alachua County Sustainability Projects Ad Hoc Committee published a report and status of sustainability in Alachua County Government, with a list of sustainable initiatives (Alachua County 2000). Then in 2008, Alachua County leaders adopted the declaration striving for the county operations to be more resilient and resource-efficient. A more specific set of resilience quantitative and qualitative goals included the goals of reducing the community-wide use of liquid fuels, including:

- Increasing the vehicle occupancy and ridership by at least 25%.
- Doubling the fuel consumption efficiency of the Alachua County government fleet.
- Moving to non-fossil fuel fleets as soon as possible. Encourage commercial and private fleets to accomplish the same.
- Reducing by 1/2 of the 13,500 annual (2008) miles driven by each registered Alachua County driver.
- Decreasing the single-occupancy vehicle trips by 25%.
- Maximizing the mobility opportunities; creating infrastructure and pathways for electric vehicles; expanding public transit, telecommuting, and flexible operating hours; using carpools, ride-sharing and car-share companies; and encourage the Metropolitan Transportation Planning Organization to adopt these policies and goals (Alachua County 2008, p. 1).

As an early commitment to countywide resilience, it is a striking document. However, the county does not provide an annual update or progress in meeting the specific goals in the declaration. County remains engaged in sustainability and resilience efforts, including work on home energy efficiency and resources for a sustainable workplace (Alachua County 2021). An added benefit for the county would be a potential measurement and reporting of resilience outcomes. The next step for the county is to engage internal and external stakeholders, revisit the declaration, and start providing consistent progress updates and reports.

Measuring, Tracking, Observing, Scrutinizing, and Reporting (MTOSR) Resilience and Sustainability

The framework for strategic resilience and sustainability planning is the ability of staff in organizations to measure, track, observe, scrutinize, and report the outcomes. While plans may be outcome-based, a greater emphasis is assigned to measurable targets and results. A comprehensive strategic resilience and sustainability planning serves as a design for leaders and administrators in decision-making. By measuring or quantifying the stated goals in the plans, local government administrators have a better appreciation of budgetary and financial resources.

The strategic resilience and sustainability plans connecting to fiscal plans aid in avoiding

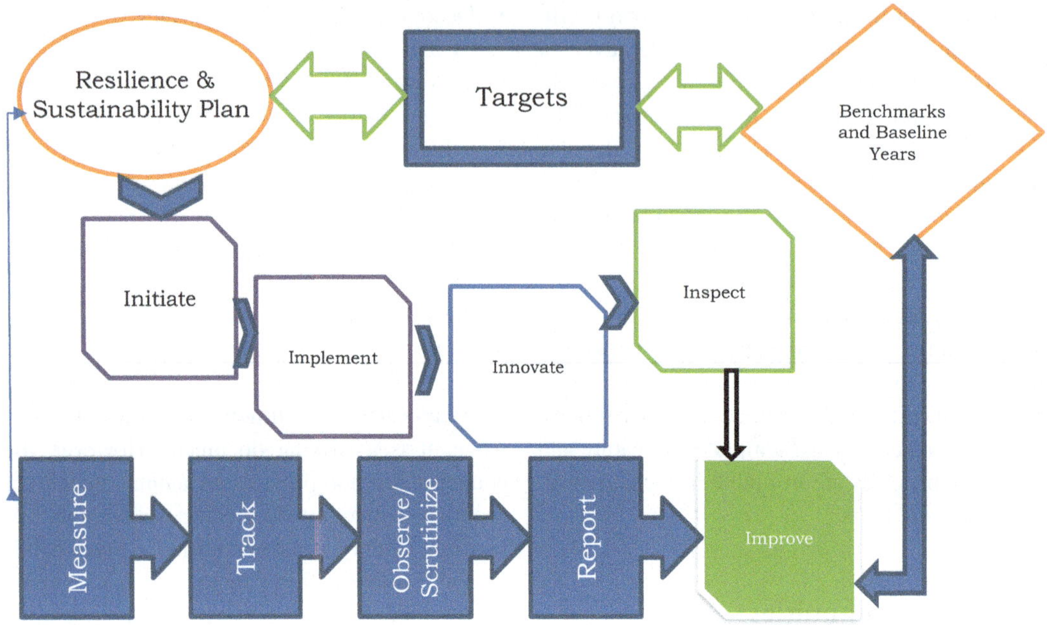

Fig. 4.1 Flowchart—measuring sustainability and resilience

the potential threats and negative effects of reduced revenues during recessions, downturns in the economy, or as in a recent case during the pandemic. Measurements of resilient and sustainable initiatives are vital for the prosperous function of an organization and improved performance. In addition to measuring economic benefits, measuring environmental, social, and governance effects of resilient and sustainable activities enable the advantageous view of the system's broad influence of activities and actions.

Cities and counties have a responsibility to track, measure, observe, scrutinize, and report results. By measuring, tracking, observing, scrutinizing, and reporting data, staff identify opportunities for cost-saving measures and cost avoidance, seek innovative solutions to problems, and deal with potential interruptions to operations to remove ambiguities and address failures. Comprehensive knowledge of costs related to all facets of facilities and fleet, equipment, and processes are instrumental to operational soundness. The costs are expressed in dollar or other currencies, and CO_2 equivalent. They measure all the activities, including the greenhouse emissions generated in the community from residential, commercial, industrial, and transportation-related activities.

Measuring the greenhouse gas emission impact is often associated with the local government's efforts to reduce a negative effect from its activities by using energy efficiency and renewable energy processes. For example, Kennedy and Sgouridis (2011) sought to define a framework for measuring the greenhouse gas emission impact from cities, labeling it as a carbon accounting "adapted to the urban scale" (p. 5260). Furthermore, the authors declared the system boundaries and the scope of the emissions for a particular city "are essential to developing a strategy for monitoring and managing urban-level carbon emissions" (Kennedy and Sgouridis 2011, p. 5268). Bader and Bleischwitz (2009) concluded how cities do not use consistent methodologies for greenhouse gas emissions, making it hard to compare to one another and standardize them.

A correct accounting procedure for greenhouse gas emission inventory includes the precisely defined system boundaries and accurate methodology for the scope of measurements. The flowchart in Fig. 4.1 entitled measuring, tracking, observing, scrutinizing, and reporting resilience and sustainability effort provides a visual representation of the several facets and characteristics of an operative reporting mechanism of the local governments' activities and initiatives. There is an explicit association between the ability to measure sustainability, resilience, and good governance. Hemmati and Enayati (2002) described good governance as "a core concept" and one that "is indispensable for building peaceful, prosperous, and democratic societies," demanding "consent and participation" (p. 41).

Good governance and answerability create a positive return on investment due to the greater acceptance of the resilience and sustainability practices within the organization and the community. The themes of answerability, good governance, and accountability succinctly link to reporting and measuring sustainability and resilience.

Key Features and Elements of Measuring, Tracking, Observing, Scrutinizing, and Reporting (MTOSR) Resilience and Sustainability

1. Establishing Goals, Vision, Mission, and Targets.
2. Determining Benchmark Data, Year, and End Year.
3. Aligning the Targets with Budget and Fiscal Plans.
4. Assigning Outcome and Target Champions.
5. Measuring, Tracking, Observing, Scrutinizing: Using Qualitative and Quantitative Data.
6. Reporting and Disseminating Results and Progress Update using the 5Is approach—Initiate, Implement, Innovate, Inspect, and Improve.

Reasons for Measuring Resilience and Sustainability Outcomes

As a result of their direct and indirect activities, organizations cause the governance, environmental, economic, and social community-wide impact. Local governments measure an indirect and direct carbon footprint and economic and social activities for improvements, and corrective measures as necessary. As an imperative, the municipal organizations measure the impact using benchmarks, baseline data, and targets and align them with their budgetary goals and keep the stakeholders and constituents informed.

The local government's role is to deliver services and quality of life outcomes. There is a demand for resilience and sustainability outcomes at the local government level and in society. Conventionally, municipalities are most concerned with their annual budgeting and service delivery. With strategic resilience and sustainability planning, local governments focus on the economic, social, environmental, and good governance goals and impacts. Similarly, it is in the local governments' interest to keep an accurate and detailed record of programs and activities' resilience and sustainability outcomes.

Over 500 years ago, in describing how to keep the accurate inventory record, Pacioli (1494, p. 11) recommended no details be omitted from the reports, adding the questions of "who, what, why, how, when, and where need to be answered." Likewise, Elkington (2012) offered a five-stage model as the critical determinant for identifying sustainability measurements toward *Zero Footprint Economy*, including the "population growth, pandemics, poverty, pollution, and the proliferation of weapons of mass destruction" (p. 13). On the other hand, Bell and Morse (2008) argued that sustainability is not measurable and cannot be defined by parameters. In their report, Renschler et al. (2010) recognized the need for more details on the PEOPLES framework they proposed to define "resilience at the community level, considering multiple dimensions and their interactions" (p. 45). The PEOPLES framework stands for Population, Environmental, Organizational, Physical, Lifestyle, Economic, Social/Cultural, attempting "to identify and group all the important components with common roles in the community welfare" (Renschler et al. 2010, p. 45).

Organizations measure resilience and sustainability outcomes for accountability and transparency and validate vision, statements, goals, and objectives. An organization measuring resilience and sustainability-related efforts proficiently conveys its characterization of sustainability, operationalizes it, and embeds it into its operational framework. In operationalizing sustainability and resilience metrics, a robust method for generating feedback and information sharing is developed. Through internal and external networks, the metric is collaboratively developed and regularly disseminated in an understandable way to all the stakeholders and collaborators.

Local governments utilize qualitative and quantitative measurements in organizational systems to seek feedback and preemptively avoid disruptions in operations. Based on a clear vision and outcomes, quantitative and qualitative data and information are monitored to measure capacity and delivery of results in any aspect of service delivery. Qualitative data documents the hierarchy of decision-making and the corresponding functions. Quantitative data is used to evaluate the functionalities of the administrative process. Measurements reaffirm the communities' commitment to resilience and sustainability programs.

Quadruple Bottom Line Measurements

The common approach to measuring resilience and sustainability outcomes in local governments is the concept of Triple Bottom Line (TBL), defined through economic, social, and environmental elements. In general, it is impractical to place each target into a single silo, as most targets belong to more than a single category. Often, the structurally essential elements of resilience and sustainability planning such as transparency, accountability, community engagements and good governance are not considered in reports and measured by local governments.

Conclusively, Alibašić (2017a, b) described resilience and sustainability planning by focusing on the Quadruple Bottom Line (QBL) and considerations of the capacity of organizations "to embed and incorporate a set of definitive policies and programs to address economic, social, environmental, and governance aspects of sustainability" (p. 41). Furthermore, governance contains the components of fiscal responsibility, participation, community engagement, transparency, answerability, moral obligation, ethical and administrative responsibility, and accountability (Alibašić 2017a, 2018, 2020).

The Quadruple Bottom Line framework expands the definition of measurements for local governments. This expanded view of responsibilities is consistent with integrating governance into measurement from moral obligation and administrative responsibility perspectives. Alibašić (2017a) allows further review and analysis of the ethics of climate resilience by integrating and expanding upon economic, environmental, social, and governance factors. The framing of climate resilience around sound economic growth, social policies, environmental protection, and good governance is consistent with the concepts of administrative responsibility and moral obligation to confront threats in the community.

For illustrative purposes, local government administrators may group targets under the Economic, Social, Environmental, and Governance themes. By adding Governance, cities further good governance, answerability, accountability, fostering accountability and transparency in their operations. The resilience and sustainability plan also draws from other plans the city had in place to align them with Goals, Outcomes, and Themes.

The Quadruple Bottom Line framework provides the overarching elements of resilience and sustainability in the plan: Economic, Social, Environmental, and Governance. For illustrative purposes, specific themes of this plan may be housed under one of the four Quadruple Bottom Line functions and specific targets are categorized under separate goals and outcomes, directly connected to other city's plans:

- Resilient Economy
- Resilient Neighborhoods
- Social Equity and Fairness
- Resilient and Safe Community
- Resilient Systems
- Resilient Transportation
- Sustainable Assets
- Fiscal Resilience
- Transparency and Accessibility
- Good Governance

The advantage of the Quadruple Bottom Line approach to resilience and sustainability planning is in using targets within the well-defined objectives and goals, benchmarking with baseline years and reporting of the results. Moreover, in identifying and assigning target champions, ownership of the outcomes in strategic resilience and sustainability plan is guaranteed.

Furthermore, all the city's climate change components, adaptation, mitigation, and climate threats are included in the resilience and sustainability plan. The new threats and responses to such dangers, such as the recent COVID-19 pandemic, are accounted for and directly connected to the emergency management and disaster preparedness plans. Qualitative and quantitative targets from a Strategic Resilience and Sustainability may include the following elements:

1. Create incentives for attraction and retention of businesses.
2. Increase the diversity and resilience of neighborhood business districts.
3. Ensure that percentage of jobs created or retained with incentives will be permanent, full-time employment with full benefits.
4. Increase incentives to the private business investments by $ (amount) millions between YEAR [2022] and Year [2026].
5. Achieve 100% energy use from renewable sources such as wind, solar, and geothermal in all city operations by YEAR [2030].

An illustrative template of the Themes, Goals, Outcomes, and Targets for the city's resilience and sustainability plan is offered below. The template breaks down areas that organizations and communities measure to improve strategic plan deliverables and outcomes.

Theme 5: Resilient Systems

- **GOAL 1:**
- **Resilient Energy Management and Greenhouse Gas Emissions Reductions**
  - **OUTCOME 1.1:**
  - **Implement initiatives to offset the effects of greenhouse gas emissions to achieve a cleaner and greener community.**

    TARGETS:
    - Reduce the City's greenhouse gas (GHG) emissions to 30% below 2010 levels by 2035.
    - Achieve 100% of the energy used in all city's operations from renewable sources such as wind, solar, and geothermal by the end of 2030.
    - Double the water reuse and recovery by the end of 2025 (from 420,000 to 840,000 gallons/day).
    - Ensure that the construction, conversion, and renovation of affordable housing align with goals to reduce greenhouse gas emissions in the City.
- **GOAL 2:**
- **Strengthen Climate Protection and Resilience**
  - **OUTCOME 2.1:**
  - **Integrate operations and preparedness measures into City plans to respond to climate change-related threats and disasters.**

    TARGETS:
    - Implement climate resilience in infrastructure planning using climate projections and incorporating climate adaptation into capital, operating and maintenance programs by the end of 2022.

        - Expand opportunities for students to learn about and take action on climate change by hiring ten interns annually.
        - Implement 100% onsite stormwater management to all new infrastructure plans by 2025.
- **GOAL 3:**
- **Resilient Reuse and Recycling Operations**

  - **OUTCOME 3.1:**
  - **Expand reuse and recycling opportunities, and composting of yard waste to decrease the amount of waste sent to landfills**

    TARGETS:
    - Reduce the number of landfill contributions by 50% within 5 years.
    - Implement a recycling campaign for all city departments and survey employees about recycling trends by 2022.

Theme 10: Good Governance

- **GOAL 1:**

- **Resilient Service Delivery**
  - **OUTCOME 1.1:**
  - Implement decisions and adapt to processes for the best use of resources to serve the needs of the entire community.

    TARGETS:
    - Increase the use of online permitting by an additional 55% over FY2020 numbers by 2024.
    - Respond to 100% of service requests for street maintenance from citizens, including potholes, Capital Improvement Projects, and rehabilitation of infrastructure within 48 hours.
    - Respond to 100% of street lighting outages within 48 hours of being reported.
    - Repair the emergency water main breaks and restore service within 4 hours 99% of the time.
    - Ensure that 100% of sidewalk snow removal complaints are abated within 24 hours of the non-compliance notice expiration.
    - Increase the City's overall Fire Code inspection completion rate to 95% by 2023.
    - Increase the conversion of 311 phone-walk-in services to digital self-serve by an additional 70% by 2023.
    - Decrease the number of walk-in customers by 70% to customer payments to the automated payment system.

Resilient Communities and Organizations Spotlight: The Quadruple Bottom Line and Resilience and Sustainability Planning in Australian Cities

The Quadruple Bottom Line method simplifies the outcomes and provides for broad goals, objectives, and measurements. Similarly, resilience and sustainability plan progress reports follows the well-defined process and identifies the planning structure, contributing to a broader acceptance and recognition of the plan. Several local governments in Australia, including Liverpool, Stirling, Penrith, Norwood, and Lake Macquarie, have adopted a Quadruple Bottom Line in their policies or programs.

For instance, the city of Joondalup depicts sustainability as meeting the challenge of striving simultaneously toward:

- Social responsibility—making decisions that lead to greater physical, cultural, and financial access and equity in service delivery and activities
- Environmental responsibility—not using more resources than required to deliver activities and services
- Economic responsibility—promoting and maintaining a city's economic development and growth in a sustainable manner

- Ethical responsibility—good governance, values, and behaviors (the City of Joondalup n.d., p. 4; City of Joondalup 2018).

Moreover, the city of Stirling administration and leadership are dedicated to "systematically review its internal policies, procedures, processes, and practices to build the organization's capacity further to deliver an ongoing quadruple bottom line (social, environmental, economic, governance) performance improvement" (City of Stirling 2009). In the city of Liverpool, the fourth bottom line is the governance of the integrated planning model, contained in the definition involving civic leadership and decision-making (Liverpool City Council 2012; n.d.). A more in-depth reporting mechanism further enhances sustainability planning in these cities. Herriman et al. (2012) discussed the essence of sustainability planning in Australian cities.

The City of Melbourne in Australia adopted the resilience strategy. It focuses on the major stressors and shocks the city faces, such as increasing population growth, climate change, unemployment, violence, natural assets, floods, bushfires, and responding to such threats (City of Melbourne 2016). The City provides an annual report on its accomplishments. The annual report published by the Resilient Melbourne office divides action items acknowledged in the resilient strategy into four distinct categories: Adapt, Survive, Thrive, and Embed. The visual representation in Fig. 4.2 shows the progress made in each category annually.

Objective Resilience and Sustainability Measurements

Resilience and sustainability activities are viewed through a top-down lens, coupled with bottom-up participation for meeting the organizational objectives and goals. Effectual stakeholder engagement is accomplished with staff being part of developing objective measurements and ultimately becoming champions for resilience and sustainability outcomes. With metrics and adequate staff engagement, employees appreciate their roles in the programs and initiatives undertaken by an organization.

Through proactive employees' engagement, measurements provide vital stakeholders and collaborators with the critical information related to goals, achievements, and the fulfillment of an organization-wide vision. It is advantageous to associate the long-term vision and the organization's commitment toward resilience and sustainability. Having the visioning and planning components expressed separately creates a sense of disconnect and barriers to implementing resilience and sustainability objectives.

In observing resilience and sustainability outcomes, organizations inform the decision-makers and employees of the current operational capacity and performance and highlight areas needing improvement, corrections, and adjustments.

In implementing measurement systems within organizations, employees receive timely and essential feedback on progress and areas needing further revisions and improvements. However, in using effective measurement techniques, organizations share success stories and adjust expectations, accept failures, and learn to adapt to changing pressures. Conversely, demonstrating quantifiable successes toward resilience and sustainability presents an opportunity for positive encouragements in the workplace, leading to higher employee engagement, and ultimately greater job satisfaction.

Objective measurements include targets, scorecards, metrics, baseline year, benchmarks, and a resilience and sustainability report. Targets are the most impartial ways to measure the outcomes of the initiatives. Targets represent a measure of collective efforts. Furthermore, targets are indicators of the organization's goals for proficient service delivery and organizational efficiency compared to baseline year using benchmarks. While service delivery is the outcome of local government activities, organizational efficiency improves service delivery, reduces operational costs, and avoids potential fiscal and budgetary pitfalls. For example, a local government may establish an energy efficiency

Fig. 4.2 Action and implementation timelines and delivery status overview of resilient Melbourne progress (City of Melbourne 2019, p. 15)

target, coupled with a renewable energy target, displaying a reduction in energy consumption and reinvesting savings into renewables. By reducing total energy consumption, the local governments can meet their renewable energy target sooner and lower costs.

Additionally, the local government organizations create and display the reduced carbon footprint due to reducing energy consumption and renewable energy production or power purchase agreement. Showing the dedication to reducing energy consumption and production or purchasing renewable energy affords the city staff to adequately reach an exact goal and target expressed in the plan. Efficiency targets may be expressed in several ways. The energy conservation outcomes are articulated as an amount of energy-reduced compared to prior years, with a baseline year over a period of time, featured in kilowatt-hours (kWh) or percentages.

Moreover, the particular energy targets are reported as savings in the amount of electricity saved or cost avoided. The targets for reducing energy consumption shown in currency/dollar amounts saved are adjusted for annual energy cost increases. Renewable energy targets may be expressed in percentages of the total annual energy consumption for all the city operations, including facilities and processes. The renewable energy target may offset the existing renewable energy portfolio in the system as states adopt the statewide renewable energy portfolio standards (RPS) (Alibašić 2017b). In some states with renewable energy portfolio standards, each power producer is required to produce a certain amount of megawatt-hours (MWh) from renewables, allowing each power user to consume a portion of renewable energy in the portfolio of energy used.

Alternatively, the renewable energy target could be expressed in kWh (kilowatt-hours), with the organization's detailed kWh to achieve annually. Both renewable energy and energy efficiency targets are tracked and observed using a baseline year and a targeted year in the future. A precise goal can be traced using a calendar or budget year. However, as most local governments have a mismatched calendar and fiscal year, an ideal reporting mechanism for resilience and sustainability strategic planning would align it with the fiscal plan calendar. Under ideal planning scenarios, each target would be attached to a budgetary process and be directly connected to the fiscal plan (Table 4.1).

Finally, empowering staff to take ownership of targets and outcomes increases the collaboration and leadership among employees. The targets created organically, owned, and championed by employees, and connected to the fiscal plan beget a greater chance of attaining and realizing desired outcomes. Each champion of the resilience and sustainability outcome and target works closely with counterparts in other departments to ensure the progress and ultimate attainment of targets and achievement of the outcomes. The facilitation and the involvement of city departments at different organizational levels foster the embedment of resilience and sustainability within an organization.

Practical Application: Resilient Energy Management Portfolio

A vital element of resilience and sustainability planning is aligned with the local government's energy efficiency and conservation initiatives. The history of energy conservation in cities and counties dates back to the first facility audits, implementing basic cost reduction strategies and evolving into modern energy efficiency strategies. Energy conservation activities remain a centerpiece of resilient cities and communities, as they are manageable, easy to measure, track, substantiate, and demonstrate an immediate return on investment.

Numerous organizations have a renewable energy target, whether to produce or procure green energy. Cities and counties commit employees to seek innovative solutions for reducing energy consumption and reinvesting energy savings into renewable energy projects. To illustrate the commitment to sustainability and to increase resilience through reduced energy consumption, one of the initial steps within organizations is to track, measure, and report the results of resilient and sustainable

Table 4.1 Creating a specific target for resilience and sustainability plan

2025 Renewable energy target	2030 Renewable energy target	2035 Renewable energy target	2025 Energy efficiency target	2030 Energy efficiency target	2035 Energy efficiency target
75% of the total energy use	90% of the total energy use	100%	Reduce consumption by 1% over the 2022 baseline year	Reduce energy use by 5% over 2022 baseline year	Reduce 10% over the 2025 energy consumption

Table 4.2 A sample of energy efficiency savings in dollars and CO_2 equivalent

2020 (kWh)	2021 (kWh)	Annual energy saved (kWh)	Cost avoidance at Avg. $.10/ kWh	CO_2 equivalents saved in metric tons
115,000,000	111,000,000	4,000,000	400,000	2,835

energy initiatives and programs. By establishing the inventory of electricity use for all the city-owned buildings, employees discern the effects of energy use on the operations. After the measurement and report of data, administrators discover potential opportunities for cost-saving projects.

As an example of reporting on energy efficiency efforts, cities and counties elected and appointed officials demonstrate meaningful energy reduction, expressed as either direct savings or cost avoidance. Investments made in energy efficiency improvements provide an obvious payback. Furthermore, the organizational leaders and administrators highlight the CO_2 equivalent removed in operations due to energy efficiency improvements and investments in renewables.

A sample of energy consumption reduction, from the baseline year of 2020–2021, benchmarked against the previous year's data, is featured in Table 4.2, with a decrease of 4,000,000 kWh. The effects of resilience and sustainability initiatives are immediately apparent. The indirect or direct savings are expressed in dollars and carbon reduction. Cost avoidance is a preferred term for local government administrators. It allows cities to reinvest the savings in energy efficiency improvements or renewable energy projects. Cost avoidance signals that ongoing expenses are continually accounted for in the present and future operations.

Using a greenhouse gas emissions calculator available at the Environmental Protection Agency (EPA) website, administrators calculate the CO_2 emission reductions and highlight the organizational commitment to mitigating climate change by reducing carbon footprint. In the Table 4.2 example, 2,835 metric tons of carbon avoided from being released into the atmosphere from energy consumption is equal to 616 passenger vehicles driven for a year or 314 of homes powered for a year (The United States Environmental Protection Agency 2021). Reports also feature successes, share developing practices, sustainability applications, savings, and provide information about future steps and projects.

By measuring, tracking, observing, scrutinizing, and reporting out the progress, administrators are accountable and garner support from constituents for the resilience and sustainability initiatives and projects. Finally, Table 4.3 provides examples of the energy projects with cost avoidance, energy savings, and environmental carbon footprint reductions. Table 4.4 features a sample of benefits from renewable energy investments. The local government can break down the report into more details, showing the sources of renewable energy. Furthermore, organizations measure, track, observe, monitor, scrutinize, and then report each energy efficiency target, renewable energy, or each project separately and report it in the annual report as overall progress toward meeting a specific energy target.

Table 4.3 Sample of energy project, with cost avoidance, energy savings, and environmental carbon footprint reductions

Type of project	Cost	Annual utility savings (kWh)	Total cost avoidance/ savings	Greenhouse gas emissions reduced (metric tons)
Window and door replacement	$1,200,000	251,911	$39,975	179
Lighting replacement	$330,000	407,337	$34,623	289
Occupancy sensors	$37,750	86,822	$7,379	62
Geothermal projects	$300,000	266,000	$22,610	189

Table 4.4 Reporting renewable energy purchase in annual reports

% of renewable energy purchased (total power)	kWh/year	Hydro	Wind	Solar	CO_2 equivalents removed through green energy purchasing (metric tons)
28%	25,948,200	74%	20%	6%	18,415

Resilient Communities and Organizations Spotlight: Climate Change and Pandemic Resilient Ann Arbor, Michigan

City of Ann Arbor, Michigan positioned itself as a leader in the combat against the threats of climate change. According to the most recent Census data, city of Ann Arbor has over 120,000 residents (United States Census Bureau 2021a, b). Moreover, the city emphasized resilience to the COVID-19 pandemic and continues to implement climate change-related initiatives (City of Ann Arbor 2021). In 2019, following the Washtenaw County Board of Commissioners' decision to unanimously endorse a climate emergency declaration, the City of Ann Arbor passed the Climate Emergency Declaration, committing itself to 100% renewable energy target by 2030 and significant climate action plan (Ross 2019; Stanton 2020). The local and regional resources were mobilized to combat climate change threats. Despite the global pandemic in 2020, in the Welcome Letter of the Ann Arbor's Living Carbon Neutrality Plan, Mayor Taylor of Ann Arbor recognized climate change as the greatest threat, reaffirming the commitment to meeting city-wide climate action targets (City of Ann Arbor 2020).

The A²Zero Carbon Neutrality Plan (City of Ann Arbor 2020, p. 6) has six core strategies:

1. Powering the city's electrical grid with 100% renewable energy;
2. Switching appliances and vehicles from gasoline, diesel, propane, coal, and natural gas to electric;
3. Improving the energy efficiency in residential homes, businesses, schools, places of worship, recreational sites, and government facilities;
4. Reducing the miles traveled in vehicles by at least 50%;
5. Changing the way of using, reusing, and disposing of materials;
6. Enhancing the resilience of people and place.

Other actions include Advancing Equity Programs, Sustaining Ann Arbor Together Grant Program, Establishing Internal Carbon Price and Offsetting Greenhouse Gas Emissions. A heavy emphasis is on carbon emissions, and "to achieve carbon neutrality, the entire Ann Arbor community must eliminate 2.1 million metric tons of carbon dioxide equivalent emissions," based on 2018 data (City of Ann Arbor 2020).

The City of Ann Arbor's approach showcases additional co-benefits from the resilience strategies from supporting the local economy, producing local energy, improving air quality, creating jobs and developing the workforce, improving local resilience, accruing cost-saving, improving public health, supporting biodiversity preservation, benefiting the most vulnerable, and scaling and transferring of plans to other communities (City of Ann Arbor 2020).

The city's interactive website offers a comprehensive overview of the sustainability progress supported by engaging and interactive tools. The website is divided into three sections, Sustainability and Me, Sustainability in the Community, and Sustainability and Natural resources (the City of Ann Arbor n.d.). Those interested in more details can view each indicator with specific statistics, data, and supplementary information for the related sustainability initiatives. It is a visually attractive system and an entertaining method to engaging the viewers with the information, including information and resources concerning COVID-19 to strengthen resilience during the pandemic.

Examples of the quantifiable goals in the strategy are featured under Action Summary Table shown in Fig. 4.3 (City of Ann Arbor 2020, p. 10). Under Strategy one, Community Choice Aggregation, which allows the city to buy power on behalf of residents and businesses, the total cost is **$3,245,000**. The greenhouse gas emissions reduced are 784,000, representing almost 36% of the total emissions, at $4/ton. The co-benefits are also expressed as better air quality, equality, and scalability (City of Ann Arbor 2020, p. 10).

Goals and indicators are identified, and measurements are in place. Most notable in the case of Ann Arbor is that each target shows the exact cost and timeline for accomplishing the desired outcomes. Short of directly determining the target champions, the city staff associated the indicators and activities with the documents in the city's operational strategies, such as plans or budget goals, and are assigned to meet the plan's specified goal and objectives. While more descriptive, these goals and corresponding indicators serve as strategies to achieve particular outcomes. Visual representation conveys the commitment of local government, elected and appointed officials to a resilient and sustainable community. The costs and benefits are explained and accounted for in the report.

Figure 4.4 communicates the cost and potential greenhouse gas emission reductions resulting from achieving the goal by 2027.

Reporting

Local governments implementing resilience and sustainability initiatives and reporting the final and interim results incorporate organizational goals and objectives. By presenting outcomes from the final implementation aspects, a local government organization communicates to constituents what it stands for, what needs to be improved, and the plans. Staff collect data and then convert the data into tables, graphs, and charts to communicate results. Benchmarking results with prior years allows for goal setting for future years.

Resilient Communities and Organizations Spotlight: Practical Application in the City of Fort Collins, Colorado Progress Report

The City of Fort Collins, Colorado, with approximate population of 170,000 people (United States Census Bureau 2021a, b) offers an effective annual sustainability progress report. The City of Fort Collins (2019a) Municipal Sustainability and Adaptation Plan features six goals.

- Resilient city operations ready to adapt to climate change and disruptive events of all scales.
- Thriving public lands supporting a healthy ecosystem and resilient to climate and population growth pressures.
- Smart water, sustainably managing water resources, and lead Colorado in water efficiency.
- Zero-waste, responsibly managing goods, products, and services throughout the life cycle to achieve waste reduction outcomes.
- Carbon-neutral transportation systems, energy production, and facilities leading the nation in sustainable performance and resource efficiency.
- A world-class workplace, a high-performing and resilient organization with a culture of sustainability.

ACTION SUMMARY TABLE

STRATEGY 1	Total Costs	GHG Reduction	%total emission	$/ton	Co-Benefits
Community Choice Aggregation	$3,245,000	784,000	35.8%	$4	AIR; $$; EQU; SCALE
Bulk Purchase of Renewables	$605,000	85,000	3.9%	$7	LOCAL; NRG; AIR; JOBS; RES; $$; SCALE
Community Solar Program	$205,000	11,500	0.5%	$18	LOCAL; NRG; AIR; JOBS; RES; $$; EQU; SCALE
Landfill Solar Project	$80,000	23,000	1%	$3.5	LOCAL; NRG; AIR; SCALE

STRATEGY 2	Total Costs	GHG Reduction	%total emission	$/ton	Co-Benefits
Home & Business Electrification	$7,100,000	362,200	16.5%	$ 20	LOCAL; AIR; JOBS; HEALTH; SCALE
Bus Electrification	$86,000,000	13,800	0.6%	$ 5,839	LOCAL; AIR; RES; HEALTH; EQU
Bulk Purchase EVs	$700,000	122,900	5.6%	$6	AIR; JOBS; RES; HEALTH; $$; SCALE
Private EV Fleets	$123,000	Not Calculated			AIR; RES; $$; HEALTH; SCALE
City EV Fleet	$4,000,000	1,100	0.05%	$3,636	AIR; JOBS; RES; HEALTH; $$; SCALE
EV Infrastructure	$42,000,000	Not Calculated			AIR; JOBS; HEALTH; SCALE

STRATEGY 3	Total Costs	GHG Reduction	%total emission	$/ton	Co-Benefits
Bulk Purchase of Energy Efficiency	$950,000	242,500	11.1%	$ 4	LOCAL; AIR; JOBS; HEALTH; $$; EQU; SCALE
Building Code Requirements	$1,935,000	48,300	2.2%	$ 40	LOCAL; JOBS; RES; $$; EQU; SCALE
LED Lighting	$3,140,000	2,600	.1%	$1,208	AIR; $$
Benchmarking	$2,370,000	Not Calculated			JOBS; $$; SCALE
Loan Loss Reserve	$1,578,000	Not Calculated			LOCAL; JOBS; RES; HEALTH; $$; EQU; SCALE
Energy Concierge & Community Education	$820,000	Not Calculated			JOBS; $$; EQU; SCALE
Net Zero Energy Affordable Housing	$800,000	400	0.0%	$2,000	LOCAL; NRG; AIR; JOBS; RES; HEALTH; $$; EQU
Green Rental Housing Program	$530,000	Not Calculated			LOCAL; AIR; JOBS; RES; HEALTH; $$; EQU; SCALE
Green Business Challenge	$700,000	Not Calculated			LOCAL; NRG; AIR; JOBS; RES; $$; SCALE
Aging in Place Efficiently	$155,000	Not Calculated			LOCAL; JOBS; RES; HEALTH; $$; EQU
Weatherization Expansion	$1,550,000	Not Calculated			LOCAL; JOBS; RES; HEALTH; $$; EQU

KEY

EQU	Benefits the most vulnerable	RES	Improves local resilience	LOCAL	Produces local energy
COST	Cost savings accrued	HEALTH	Improves public health	ENV	Supports biodiversity preservation
$$	Supports local economy	SCALE	Scalable or transferable to other communities		
AIR	Improves air quality	JOBS	Creates jobs and/or develops workforce		

Fig. 4.3 Summary of strategies, actions, cost, and co-benefits for proposed actions in the A^{2}Zero plan (City of Ann Arbor 2020, p. 10)

Under goal number 5 of carbon neutrality, the City of Fort Collins (2019a) strives to be carbon neutral by 2050 and plans to have the "transportation systems, energy production and facilities lead the nation in sustainable performance and resource efficiency" (p. 28). Moreover, the City

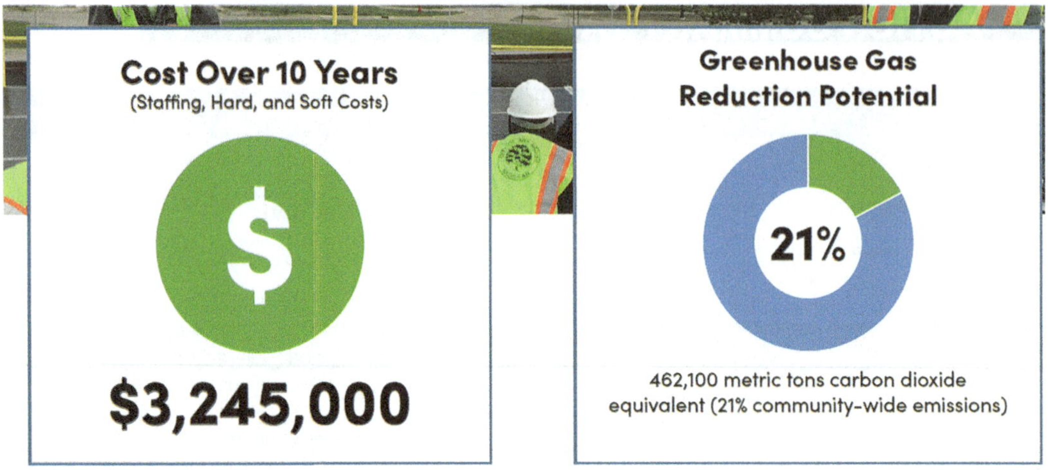

Fig. 4.4 City of Ann Arbor (2020, p. 23). ACTION 1: Implementing community choice aggregation

of Fort Collins (2019a) "uses the same incremental goals as the community of a 20% reduction in greenhouse gas emissions from the 2005 baseline by 2020 and 80% by 2030," and "in 2017, the City exceeded its 2020 goal by reducing emissions by 21%" (p. 28). The city also has a 100% renewable energy goal.

The city publishes an annual carbon inventory report showing progress toward carbon neutrality. The City of Fort Collins (2018, 2019b) reported that on an organizational level, it was leading in the community in "greenhouse gas reduction goals: 20% reduction below 2005 levels by 2020, 80% reduction by 2030, and carbon neutral by 2050," adding that "as of 2018 the City's municipal inventory is 22% below 2005 levels." The year 2005 serves as a baseline year for the City. Since 2004, the city sets specific standards for annual reporting and measurements, and importantly for observing, scrutinizing, and monitoring progress (City of Fort Collins 2004; City of Fort Collins 2006).

Reporting is a crucial piece of the sustainability and resilience planning. With correct reporting, local governments feature results, are transparent and accountable to the local constituency. A notable component of the cities' engagement in sustainability and resilience is attributed to the sustainability plan and annual reporting. The local government plans serve as a multiyear, adaptable blueprint used by each department to plan activities and justify budget requests based on economic, social, environmental, and governance sustainability-related outcomes.

Centralization of Reporting

The local governments report total progress in all areas of resilience and sustainability by showing how many specific targets or indicators have been met, how many are making progress, or not making progress at all. Overall results may indicate the total percentage of the outcomes and targets met of the total number of targets adopted and outcomes achieved by the local government, expressed in percentages or total numbers. Furthermore, these targets and outcomes can be compared to a previous year's numbers and can be broken down to a theme, goal, or objective.

A review of the diverse resilience plans, sustainability plans, policies, and programs in local governments revealed the importance of having a point of contact for measuring, tracking, observing, scrutinizing, and reporting results and outcomes. In Ann Arbor, it is the Office of Sustainability and Innovations; the Office of Climate Action, Sustainability, and Resiliency in Denver, CO; the Sustainability Services in Fort Collins, CO; and Sustainability and Resilience Division in Tallahassee (City of Ann Arbor n.d.; City of Denver 2021; City of Fort Collins n.d.;

City of Tallahassee 2021). A central and unified office with staff is in place to compile the collected reports, making the efforts more effective. While cities resilience initiatives differ from community to community in nature and outcomes, they reveal the relevance of consistency and ownership of the process of reporting the resilience and sustainability activities.

Having a central point of contact for data collection and reporting of all the resilience and sustainability actions and activities enables staff to continue contributing to sustainability and resilience and, at the same time, focus on completing their daily tasks. A strategic resilience and sustainability plan becomes an integral part of the budgeting process by pursuing resilience and sustainability goals, objectives, outcomes, and targets.

Cities have an enormous impact on the economy, society, and the environment through decisive actions to reduce energy demand and energy consumption. Local governments make day-to-day crucial decisions to provide services and meet increased demand for services while facing constant and severe budget constraints to staffing and operations. At the same time, the city governments are expected to provide the same level of services without additional revenues or resources. Institutionalizing resilience and sustainability is a colossal undertaking, requiring leadership and readiness to measure, track, observe, monitor, scrutinize, and report progress. Tracking, measuring, observing, scrutinizing, and reporting data are crucial for all projects.

Each department collects and disseminates data through a single point of contact, utilizing a single person in the centralized resilience and sustainability office. With reporting, the department is still responsible for collecting data. However, the reports are done on a macro-level to reveal the city's overall results noticeably. When available funding is in peril, the local government's capacity to pursue resilience and sustainability objectives becomes difficult, as the cities provide essential services. Resilience planning becomes an integral part of the budgeting process through the active pursuit of resilience and sustainability goals and targets using the Quadruple Bottom Line approach to reporting (Table 4.5).

Table 4.5 Centralized reporting: framework for resilience and sustainability progress reports

Qaudruple bottom line (QBL) benefits	Centralized reporting
Carbon reduction/ environmental	Health Benefits, reduced pollution, reduced greenhouse gas emissions/CO_2 emissions reductions, clean rivers, improved water quality, regional air quality benefits, climate change, clean energy jobs, biodiversity.
Resilient systems/societal	Sustainable operations, clean energy, energy change, resilience, resilient communities and organizations, diverse population.
Resilient operations and organization/ governance	Cost cutting, cost avoidance, price reduction, kWh reduced, savings, frugality, answerability, transparency, good governance, stewardship, Return on Investment, accountability, responsibility, and responsiveness, budget alignment, fiscal resiliency, diverse workforce, equity, ethics, justice, fairness.
Resilient economy/ economic	Job creation, transformation, diversity, city attractiveness, economic growth and development, business support.

Summary

Organizations with strategic resilience and sustainability plans adapt, transform, accept disruptions and deliver services without interruptions for communities and constituents. In a crisis like the one in 2020 and 2021, resilience- and sustainability-related initiatives convert into an opportunity and a tool for counties, cities, and organizations in general to adjust priorities and outcomes of the budget process. The approach of comprehensive accountability and transparency provides the primary direction for the goals, outcomes, and targets established in a strategic plan.

The Quadruple Bottom Line approach to measuring resilience and sustainability outcomes ensures the administrators and staff are held accountable. Progress reports available online or in print improve transparency and extend an opportunity for community engagement. Cities and counties measure, track, observe, scrutinize, and report targets regularly. Measuring targets allows administrators to value the outcomes and goals being measured and tracked. Measurement allows for complete accountability and transparency of the implementation of resilience and sustainability initiatives. The design of the resilience programs does not automatically produce more efficiency unless it is coupled with tracking, measurements, observations, adjustments, and reports of the outcomes.

The foundation for resilience and sustainability vision is measuring, tracking, and reporting the initiatives, programs, and outcomes. Once a resilience target achievement is reported, measured, and then compared to various outcomes, it assists organizations in assessing the positive impact on the overall effectiveness of local government service delivery. Progress reports availability online is crucial for transparency of resilience and sustainability-related efforts. Targets showing progress reported annually are connected to an annual financial plan and budgeting processes.

Organizations operationalize the resilience and sustainability metrics for better service delivery. The resilience and sustainability activities are centralized, embedded, and directly related to the annual budget. The centralization of reporting through a single department is critical. Reporting furthers the transparency, accountability, answerability, and good governance of the local government and other organizations' operations and strengthens community resilience.

Municipal organizations use positive, practical examples of resilience-related projects to improve the overall governance of their operations and deliver better services for the community. For instance, sustainable energy initiatives function as the essential platform for organizational engagement in such operational efforts. In encouraging the ownership of resilience and sustainability targets, city employees are accountable and in control of sustainability and resilience initiatives.

Outcomes, Discussions, and Further Considerations

- Assess the importance of measuring, tracking, observing, scrutinizing, and reporting the resilience and sustainability targets and outcomes.
- Explain the points of differences between sustainability indicators and resilient and sustainable targets.
- Discuss diverse approaches the municipal organizations undertake in defining and reporting measurable outcomes and provide a sample of targets and corresponding measurements of resilience and sustainability efforts.
- Review organizational annual resilience and sustainability plan reports.
- Analyze and compare the metrics, goals, targets, baseline, sources of data, and other reporting areas.
- Discuss changes in reporting mechanisms due to pandemic and climate change threats.

References

Alachua County (2000) The Alachua County Sustainability Projects Ad Hoc Committee. Report on the status of sustainability in Alachua County Government. https://p2infohouse.org/ref/17/16999.pdf

Alachua County (2008) Declaration for resource efficient and resilient county. http://www.alachuacounty.us/Depts/Manager/

Alachua County (2021) Sustainability. https://alachuacounty.us/Depts/Sustainability/Pages/Sustainability.aspx

Alibašić H (2017a) Measuring the sustainability impact in local governments using the quadruple bottom line. Int J Sustain Policy Pract 13(3):37–45

Alibašić H (2017b) Sustainable energy policy for state and local governments. In: Farazmand A (ed) Global encyclopedia of public administration, public policy, and governance. Springer International Publishing AG, Cham. https://doi.org/10.1007/978-3-319-31816-5_3430-1

Alibašić H (2018) Sustainability and resilience planning for local governments: the quadruple bottom line strategy. Springer, New York, NY. https://www.springer.com/us/book/9783319725673

Alibašić H (2020) The administrative and ethical considerations of climate resilience: the politics and consequences of climate change. Public Integr:1–18. https://doi.org/10.1080/10999922.2020.1838142

Bader N, Bleischwitz R (2009) Measuring urban greenhouse gas emissions: the challenge of comparability. Cities Clim Change 2(3):1. https://journals.openedition.org/sapiens/854

Bell S, Morse S (2008) Sustainability indicators: measuring the immeasurable? Earthscan Publishing, Washington, DC

City of Ann Arbor (2020) Ann Arbor's living carbon neutrality plan. https://www.a2gov.org/departments/sustainability/Documents/A2Zero%20Climate%20Action%20Plan%20_3.0.pdf

City of Ann Arbor (2021) Sustainability and resilience during COVID-19. https://www.a2gov.org/departments/sustainability/Adaptation-Resilience/Pages/Sustainability-and-Resiliency-During-COVID-19.aspx

City of Ann Arbor (n.d.) Sustainability and innovations home. https://www.a2gov.org/departments/sustainability/Pages/default.aspx

City of Denver (2021) Office of climate action, sustainability, and resiliency. https://www.denvergov.org/Government/Departments/Climate-Action-Sustainability-Resiliency

City of Fort Collins (2004) Action plan for sustainability: policy and recommended strategies. https://www.fcgov.com/sustainability/files/2004-sustainabilityactionplan.pdf?1566857909

City of Fort Collins (2006) The path of sustainability: city of Fort Collins action plan for sustainability 2006 report on municipal operations - April 2007. https://www.fcgov.com/sustainability/files/2006-sustainabilityannualreport.pdf?1566857909

City of Fort Collins (2018) 2018 Municipal carbon inventory report. https://www.fcgov.com/sustainability/files/2018-carboninventoryreport.pdf?1583786004

City of Fort Collins (2019a). 2019 Municipal sustainability and adaptation plan. https://www.fcgov.com/sustainability/files/2019-sustainabilityandadaptationplan.pdf?1566857909

City of Fort Collins (2019b) 2019 Progress report. https://www.fcgov.com/climateaction/files/2019-municipal-sustainability-adaptation-report.pdf?1606159566

City of Fort Collins (n.d.) Sustainability services. https://www.fcgov.com/sustainability/

City of Joondalup (2018). Joondalup 2022: Strategic Community Plan 2012-2022 Summary Document. Retrieved from https://www.joondalup.wa.gov.au/wp-content/uploads/2018/07/Joondalup-2022-Summary-Document-Updated-July-2018.pdf

City of Joondalup (n.d.) Policy 5-4 sustainability. http://www.joondalup.wa.gov.au/Libraries/Policies/Sustainability_5-4.pdf

City of Melbourne (2016) Resilient Melbourne. https://resilientmelbourne.com.au/wp-content/uploads/2016/05/COM_SERVICE_PROD-9860726-v1-Final_Resilient_Melbourne_strategy_for_web_180516.pdf

City of Melbourne (2019) Resilient Melbourne Delivery Office annual report 2018–2019. https://resilientmelbourne.com.au/wp-content/uploads/2019/10/Resilient-Melbourne-Annual-Report-2018-19-Full-Double-Spread-Version-Web.pdf

City of Stirling (2009) Sustainability policy. https://www.stirling.wa.gov.au/your-city/documents-and-publications/your-city/about-council/governance-and-transparency/policies/sustainability-policy

City of Tallahassee (2021) Sustainability and resilience division https://www.talgov.com/publicsafety/resilience.aspx

Elkington J (2012) The Zeronauts: breaking the sustainability barrier. Routledge, New York, NY

Hemmati M, Enayati J (2002) Multi-stakeholder processes for governance and sustainability: beyond deadlock and conflict. Earthscan, London

Herriman J, Storey H, Smith P, Collier G (2012) Working relationships for sustainability: improving work-based relationships in local government to bring about sustainability goals. Commonw J Local Gov 10:116–133. http://epress.lib.uts.edu.au/journals/index.php/cjlg/article/view/2693/2906

Kennedy S, Sgouridis S (2011) Rigorous classification and carbon accounting principles for low and Zero Carbon Cities. Energy Policy 39(9):5259–5268. https://doi.org/10.1016/j.enpol.2011.05.038

Liverpool City Council (2012) Integrated environmental sustainability action plan. http://www.liverpool.nsw.gov.au/trim/documents?RecordNumber=089986.2016

Liverpool City Council (n.d.) Our vision for sustainability. http://www.liverpool.nsw.gov.au/environment/sustainability

Pacioli LF (1494) Particularis de computis et scripturis (Accounting Books and Records from From Pacioli's Summa de Arithmetica Geometria Proportioni et Proportionalita) (J. Cripps trans.). Pacioli Society, Seattle, WA

Renschler CS, Fraizer AE, Arendt LA, Cimellaro G, Reinhorn AM, Bruneau M (2010) A framework for defining and measuring resilience at the community scale: the peoples resilience framework. National Institute of Standards and Technology, Gaithersburg, MD. https://www.esf.edu/glrc/library/documents/FrameworkforDefiningandMeasuringResilience_2010.pdf

Ross M (2019) Washtenaw County declares local, global 'climate emergency'. MLive. https://www.mlive.com/news/ann-arbor/2019/09/washtenaw-county-declares-local-global-climate-emergency.html

Stanton R (2020) Ann Arbor declares climate emergency, sets 2030 carbon-neutral goal. MLive. https://www.mlive.com/news/ann-arbor/2019/11/ann-arbor-declares-climate-emergency-sets-2030-carbon-neutral-goal.html

The United States Environmental Protection Agency (EPA) (2021). Greenhouse gas equivalencies calculator. Retrieved from https://www.epa.gov/energy/greenhouse-gas-equivalencies-calculator

United States Census Bureau (2021a) Ann Arbor city, Michigan QuickFacts. https://www.census.gov/quickfacts/fact/table/annarborcitymichigan/PST045219

United States Census Bureau (2021b) Fort Collins city, Colorado. QuickFacts. https://www.census.gov/quickfacts/fact/table/fortcollinscitycolorado/PST045219

5 Implementing a Strategic Resilience and Sustainability Plan: From Policies to Initiatives, Programs, and Projects

"What one can observe in the world, however, is that neither the state nor the market is uniformly successful in enabling individuals to sustain long-term, productive use of natural resource systems. Further, communities of individuals have relied on institutions resembling neither the state nor the market to govern some resource systems with reasonable degrees of success over long periods of time."

(Ostrom 1990, p. 1).

Overview

Chapter five details the implementation of a strategic resilience and sustainability plan, describing programs and initiatives in communities to achieve resilient and sustainable organizations and communities. Furthermore, the book chapter outlines the activities, initiatives, and outcomes, the championing and implementing the strategic resilience and sustainability plan. Moreover, an extensive review of the implementation of resilient and sustainable energy practices, initiatives, and projects is provided. The four distinct domains of Quadruple Bottom Line (QBL) and corresponding implementation strategies are governance, economic, environmental, and social functions of resilient and sustainable community planning. Modern cities, townships, villages, and counties implement comprehensive, quantifiable resilience and sustainability projects and initiatives with an ultimate goal of cost and carbon reduction, a healthier environment, a more robust economy, improved social outcomes, and efficient service delivery and good governance.

Keywords

Project implementation · Sustainable energy · Energy efficiency · Renewable energy · Carbon footprint · Greenhouse gas (GHG) emissions · Resilient energy · Measuring outcomes · Carbon reduction · Waste minimization · Resilient water systems · Resilient transportation · Resilient Public Safety Services · Resilience · Good governance · Resilient communities · Resilient economy · Resilient environment · Social equity · Resilient operations

Key Questions

The chapter of this book answers the following underlying premises and inquiries:

- What are resilience- and sustainability-related activities, initiatives, and outcomes that contribute to the reduction of the greenhouse gas emissions, support economic development and growth, address social issues and promote good governance?

H. Alibašić, *Strategic Resilience and Sustainability Planning*, Sustainable Development Goals Series, https://doi.org/10.1007/978-3-030-91159-1_5

- What type of resilience and sustainability policies, programs and projects are effective at addressing climate change threats?
- How do cities decide and select particular programs and projects to implement?
- How do resilience and sustainability projects produce a return on investment?
- Who champions and implements the plans?

Introduction

The embedment of resilience and sustainability strategies at all organizational levels is a holistic, systemic, and dynamic process. The previous book chapters recounted the vital components contributing to the achievement of robust resilience and sustainability outcomes, including but not limited to:

- Internal and external examination, stakeholder and partners' identification, and operational capacity analysis.
- Meaningful stages of policies and programs toward creating and implementing a robust resilience and sustainability strategic plan.
- Internal and external stakeholder engagement and pursuance of partnership and collaborative opportunities in achieving resilient and sustainable outcomes.
- Employees empowerment to define, own, and champion resilience and resilience targets and outcomes.
- Measuring, tracking, observing, scrutinizing, and reporting targets and results and creating and sharing resilience and sustainability plan progress reports.

Organizations design projects, components, and activities to implement strategic resilience and sustainability plan initiatives and commit financially to the resilient and sustainable goals and objectives. By connecting resilience and sustainability strategic planning directly to the budget process, organizations indicate and substantiate commitment to long-term planning processes.

From Plans to Applied Resilience and Sustainability Initiatives

In recent years, local governments showcased the resolve to invest in infrastructure to design the framework and implement strategies for sustainability and resilience. Cities and counties unceasingly face economic recession, extreme weather-related exposures, disasters, pandemics, and climate change. Organizations invest preemptively into resilience and sustainability projects and programs to counter internal and external threats and pursue opportunities for improvements. The imperative in implementing resilience to counter climate change and pandemics threats is to establish an appropriate balance between the organizational objectives, vision, and effectiveness in delivering services. Municipalities may evaluate the implementation mechanism of the planning through active assessment of the investments in infrastructure improvements by asking the following questions:

- How well do the streets and road infrastructure perform under current stressors?
- How many miles of sidewalks are missing along major roads and streets?
- How many miles of new bike lanes did the local government add in recent years?
- How many miles of power lines were buried in moving power lines away from trees?
- How many trees were planted in recent years in city parks?
- How many residents and businesses did an organization engage in community outreach events and activities?
- What are the governance, economic, environmental, and social benefits from investments in resilience and sustainability initiatives?
- What are the steps to implementing low-impact design and engaging businesses to adopt emerging practices to reduce stormwater runoff?
- What benefits does an organization achieve from investments in sustainable energy, including renewables?

Local government organizations benefit from investments in renewables and sustainable energy initiatives and projects. The administrators link sustainable energy initiatives to a fiscal plan to maximize gains for the organizations. Implementing sustainable energy projects is an effective long-term strategy with governance, economic, environmental, and social advancements aligned in a systematic and purposefully defined approach. Commonly, the administrators' and managers' approach to implementing resilience and sustainability initiatives includes the following features.

Comprehensiveness, Robustness, and System and Project-driven Processes

Evaluating each project and integrating it into the system-wide plans. Together with outcome and target champions, each resilience and sustainability target links to a single project in a reporting design with projected outomes. As an illustration, in strategies to implement energy objectives, power use, or power production, each project contributes to the overall energy reduction to meet renewable energy targets and carbon footprint and greenhouse gas emissions reduction. The process leads to robustness in the implementation of the projects using a system-wide approach.

Targets Are Quantified and Budgetary and Policy Aligned

Each target is designed to be directly aligned to the annual fiscal and budgetary process and eventually revised and adjusted to existing plans and policies. The targets need to be practical and realistic to lead to the most effective outcomes.

Accountability, Transparency, and Answerability

Local governments' accountability, transparency, and answerability synergies are undeviatingly related to the organizational aptitude to implement and execute flexible, durable, resilient, and sustainable strategies, with regular reports, updates, and robust community engagement. Imperative to good governance and ethical decision making are all three elements expressed through accountability, answerability, and transparency.

The Organizational Dynamic of Project Implementation

Each organization has a unique set of goals and objectives related to sustainability and resilience. Hardjono, Van Marewijk, and de Klein (2004) conceived the European Corporate Sustainability Framework (ECSF), providing a methodology for implementation of sustainability in organizations relying on organization dynamics to address social and environmental problems (as cited by Stubbs and Cocklin 2008). The organization deploys representative interaction methodology in a systematic approach to engaging in resilience and sustainability initiatives.

The implementation of resilience and sustainability projects is accomplished by addressing all interconnected aspects of interests for the organizations, including but not limited to governance, economic growth and development, social equity and justice, and environmental concerns. The organizational dynamics are relevant to all features of implementation strategies. As organizations attempt to address projects' economic, environmental, and social issues, the often neglected portion of project implementation is good governance. Alibašić (2017) argued in favor of the local governments and other organizations to utilize the Quadruple Bottom Line (QBL) to implement resilience and sustainability strategies to improve service delivery and increase answerability, transparency, and accountability benefits from adding governance to strategic planning.

Additionally, good governance is critical in municipalities' efforts to design and manage local governments' operations with the lowest carbon footprint and impact. Managers and administrators play a precarious role in managing resources appropriately for a resilient and sustainable community. The appointed officials share common apprehension about resource management and service delivery in operational planning and long-term strategies. Municipal leaders include the following elements in strategic resilience and sustainability plans: robust and flexible economic development, sustainable energy, resilient water, natural resources, resilient transportation, equitable programs, and sustainable waste management systems. Such systems are not resource-dependent and are renewable resource-based, efficient, resistant, robust, and durable. The primary objective of organizations is the improved conservation and use of resources. Furthermore, municipal administrators and managers attempt to achieve reduced operations costs, increased efficiency, and meeting the governance, economic, environmental, and social goals for the organization and community.

The Quadruple Bottom Line (QBL) Implementation Objectives

The four distinct implementation strategies are governance, economic, environmental, and social elements of resilient and sustainable strategic planning.

Resilient Operations and Good Governance

- Approach government operations from resilient and sustainable practices, aiming for fiscal resilience, transparency, accountability, accessibility, inclusiveness, and intentionality in outreach to increase opportunities for marginalized and vulnerable populations.
- Strengthen infrastructure, improving building envelopes, mobility, and transportation options.
- Improve communication strategies to share frequent updates with the community.

Resilient and Prosperous Economy

- Advance positive economic development and growth drivers.
- Spur the local and regional economic growth opportunities by supporting job creation, updating and redeveloping neglected properties, removing the barriers to business development, and favoring environmentally responsible and socially conscious industries.
  - Attract and retain diverse entrepreneurship opportunities and long-term investments in the community businesses.

Resilient Environment and Natural Resources

- Protect water resources, rivers, tributaries, streams, lakes, and oceans.
- Reduce pollution, rehabilitate the riparian system to restore the ecosystem to a more natural state.
- Provide greater access to parks and recreational opportunities to contribute to the overall health and vitality of the community.
- Decrease energy consumption and increase renewable energy production and use.
- Minimize waste, reuse and repurpose materials, reduce consumption, promote and improve recycling.
- Seek operational methods to reduce the carbon footprint within an organization and a community.

Resilient Social Capital in Equitable Communities

- Ensure equitability, fairness, social justice, improve affordable housing, address racial and income inequalities, and remove obstacles to quality education.
- Support redevelopment, promote downtowns, and provide resilient public safety services in the neighborhood and business districts.

- Increase community engagement and participation in service delivery.
- Improve emergency preparedness and disaster responses to be robust and include climate change and pandemic responses in planning.

The Quadruple Bottom Line (QBL) implementation framework enables administrators to align projects within specific planning components. In attempting to meet the QBL objectives, organizations intentionally design the critical components of the implementation strategies with a system-wide approach. The following list provides a snapshot overview of some of the key resilience and sustainability initiatives and strategies, followed by a more in-depth review of sustainable energy and water protection strategies in the proceeding pages.

Implementation of Resilience and Sustainability Initiatives

Resilient Transportation Infrastructure and Systems

Communities require a diversified mode of transportation, accessible public transit, functional roads, and safe traffic conditions. A significant component in the system-wide assessment of the quality of life is the state of the community's facilities, infrastructure, services, and amenities, measured in the quality to meet that community's needs and expectations. For example, a road asset management plan may include the expansion of bike lanes and a reduction in the number of traffic lanes in communities.

QBL Categories: Resilient and Prosperous Economy, Resilient Environment and Natural Resources, Resilient Operations and Good Governance.

Resilient Public Safety Services

Local governments provide public safety services and have emergency preparedness and disaster mitigation plans in place with consideration for unexpected circumstances and threats, such as pandemics.

QBL Categories: Resilient Social Capital in Equitable Communities.

Carbon Footprint Reduction

Each activity identified as operational efficiency and service delivery is directly or indirectly associated with reducing greenhouse gas emissions and decreasing carbon footprint.

QBL Categories: Resilient Social Capital in Equitable Communities, Resilient and Prosperous Economy, Resilient Environment and Natural Resources, Resilient Operations and Good Governance.

Resilient Water Systems

The systems for water protection and delivery, stormwater, and sewer services are essential components in the operational strategies of municipalities and the infrastructure for service delivery. Local government elected and appointed officials are interested in the protection of water resources using multiple strategies such as reducing water waste in the water delivery system, adopting policies of ending the purchase of bottled water for events, and raising awareness about the adverse effects on water, environment, health, and economic consequences of plastic microbeads. Water assets are critical for the health and wealth of a community.

In demonstrating a commitment to environmental stewardship, asset management planning is a beneficial mechanism available to the administrators and managers. Delivering particular service levels is achieved through efficient, reliable, resilient, and sustainable management of the operating systems. Municipalities use proactive long-term planning for asset management to achieve resilient operating systems and sustainable organizations and ensure the community's wellbeing and a healthy environment for the current and future generations. Asset management includes the planning, design, construction, operation, and maintenance of infrastructure in the organization and community. For example, asset management recommends emerging practices, light infrastructure development, and stormwater management improvements to increase the water quality. Those initiatives may include bioswales, cisterns, wetlands, trees, rain gardens, rain barrels, green roofs, permeable pavement, more parks, open spaces, green streets and alleys, controlling water on-site, and other best practices.

QBL Categories: Resilient Operations and Good Governance, Resilient and Prosperous Economy, Resilient Environment and Natural Resources.

Waste Minimization, Reuse, and Recycling

Waste minimization, reduction, recycling, and repurposing of materials are some of the resilient and sustainable initiatives to lessen the negative effects on the environment stemming from production and service delivery.

QBL Categories: Resilient Social Capital in Equitable Communities, Resilient and Prosperous Economy, Resilient Environment and Natural Resources, Resilient Operations and Good Governance.

Resilient Energy

Local governments own buildings that consume power and use equipment and vehicles in operations. Administrators are interested in reducing costs of heating and cooling, consumption of fuel, and power consumption.

QBL Categories: Resilient Social Capital in Equitable Communities, Resilient and Prosperous Economy, Resilient Environment and Natural Resources, Resilient Operations and Good Governance.

Resilient Energy System Strategies

Over the past decade, cities have been increasingly active in advocating for environmental responsibility, actions to reduce carbon footprint, resilience, and sustainability. For organizations, energy efficiency is the cornerstone of resilience and sustainability attempts. The administrators pursue energy efficiency, energy management, and renewable energy programs and policies, with all three elements defined as a resilient energy strategy. With municipal budgets constrained or in crisis, cost savings are imperative. In most organizations, energy costs are a notable component of their budgets, and regardless of the geographical position of those local governments, the energy costs are increasing each year.

Organizations launched energy efficiency projects to reduce operational costs, from insulating government buildings, investing in green energy, geothermal, replacing single-pane windows, to upgrading HVAC equipment. Modeling the energy supply and demand for resilient cities requires new paradigms. Organizations adopt a more systemic and strategic Quadruple Bottom Line approach to energy efficiency and renewable energy. It assists them in achieving the most notable energy savings and integrating the energy efforts into the greenhouse gas emission goals.

The four critical actions for implementing sustainable energy include:

- Institutionalizing energy efficiency and resilience planning into the organizational culture.
- Dedicating staff to data collection and other resources to detailed measurement and progress updates and reporting.
- Developing a long-term energy efficiency, energy management, and conservation strategy to guide the energy management work.
- Empowering the staff to innovate and pursue smart energy projects.

The difficulties of changing the infrastructure for energy supply and demand on a larger scale are evident in the built environment. Most state-level policies are not designed to support local communities' implementation of sustainable energy programs, and the role cities play in structural changes of a large-scale shift to sustainable energy infrastructure and production. There are diverse energy sources available to cities and counties, including solar, wind, geothermal, and hydro. In addition to costs, the linkage between energy production and consumption to greenhouse gas emission is the key to analyzing and examining sustainable energy options.

In the analytical framework for evaluating energy policies, categorizing environmental, social, governance, and economic issues stemming from energy consumption is more complex at a community-wide level. The multiple factors include the type of buildings, residential or business, seasons, and age of building stock. For illustration, Douthat et al. (2020) examined passive building characteristics in residential energy use during the summer months. The examination is unique to each city.

Resilient Energy Plan and Quadruple Bottom Line (QBL)

Despite the obstacles and non-existent national energy strategies, coupled with a diverse patchwork of state-level energy policies, communities across the country are seeking the benefits of resilient energy programs and projects. The role of the sustainable energy plan for local governments is to identify and prioritize applications based on specific community needs, thereby leading to the implementation of the programs most likely to accomplish those goals.

Local governments' focus on sustainable energy and energy efficiency resulted in projects completed or underway shedding millions of kilowatt-hours (kWh) of electricity from their operations over the past two decades.

Further evidence of resilient communities are the efforts to capitalize on energy efficiency and sustainable energy strategies, leading to less reliance on traditional sources of energy in the energy management portfolio. Appointed officials and staff regularly and consistently analyze, evaluate, and explore cost-effective opportunities for on-site energy generation, including the use of solar panels and geothermal production technologies. The primary focus of the energy efficiency strategy for the organizations is to reduce or avoid costs in operations and improve energy management.

With the development of the energy efficiency and conservation strategy, administrators understand the carbon footprint for facilities and vehicle fleets. In some instances, organizations used it as an opportunity to collect data and develop a baseline report for all the greenhouse gas emissions generated in the community from residential, commercial, industrial, and transportation-related activities. Alibašić (2018) argued the local governments' sustainable energy goals address budgetary issues, respond to constituents' demands, and meet the resilience and sustainability targets while anticipating changes in energy markets. Municipalities use strategic resilience and sustainability planning to counter negative governance, economic, environmental, and social characteristics of the traditional energy generation and use (Alibašić 2018).

In reviewing sustainable energy options and suggesting solutions, MacKay (2009) conferred on the energy supply, carbon pollution, and the tax system and how and why society is still dependent on certain types of energy sources by comparing the cost of standard or long-established energy sources to greener or renewable energy sources, such as solar, wind, and geothermal. Analysis of diverse, sustainable energy options, costs, and benefits is critical for cities, counties, and other organizations as they consider resilient projects and policies. Sovacool and Watts (2009) maintained that the best policy approaches mixed with appropriate leadership and current technology in New Zealand, the United States, and the rest of the world could fully be powered by renewable energy.

The policymakers challenge administrators to achieve 100% renewable production. The 100% renewable energy targets have been in place for some time now. However, the number of local governments across the United States committing to the renewable energy goal has increased in recent years. Over 250 United States Conference of Mayors (USCM) members and the city councils/city commissions adopted the 100% renewable energy targets (USCM 2017). Stafford and McCann (2020) reported how in Utah alone, "23 cities and counties have resolved to adopt 100 percent net-renewable electricity by 2030, representing about 37 percent of Utah's electricity load" (para. 3). UCLA Luskin Center for Innovation (2019, p. 2) reported that "200 cities and counties now have committed to or already achieved 100% clean electricity."

The EPA regularly updates and features top 30 organizations from the private and public sector, including local governments procuring or producing renewable energy (the United States Environmental Protection Agency 2018). Alexander and Boyle (2010) argued for renewable power and evaluated various types of energy sources, and the overall worldwide impact of renewable energy sources "was already providing a significant proportion of the world's primary energy" (p. 14). Nevertheless, the United States Energy Information Administration (2019) shows that renewable energy production has doubled since 2008 in the United States. However, the overall renewable energy production at the current level is not nearly enough to slow down the negative effects of climate change.

Local governments contemplate carbon dioxide (CO_2) and the impact of human activity on climate change and damaging effects on the ecosystem and society in defining the reasonable policy approaches to energy production and use. The enormous energy use by cities has a significant climate change impact due to greenhouse gas emissions from energy used in urban settings. Cities are major contributors to greenhouse gas emissions (the United States Environmental Protection Agency 2019). However, the perceived ease of achieving a 100% renewable energy target is an issue that organizations will have to address. Kohl (2000) posited how renewable energy sources are on the rise, and "often, the potential of the various technologies which exploit renewable energy sources is regarded with skepticism" (p. 6).

Again, the ability to be in a leading position is advanced through evaluation, exploration, and implementation of the most effective renewable energy technologies. Realistically and in meeting the expectations of the constituents in their communities, administrations have undertaken the following actions to achieve the goals of sustainable and resilient energy plan and strategies for implementation:

- Develop energy conservation strategies to focus on energy efficiency and decrease energy consumption and demand throughout the organization and community.
- Set a goal to achieve an exact percentage of the city's renewable sources by a target year to diversify energy sources.
- Reduce total fuel consumption by a certain percentage over the period of time, equating it to the annual savings and carbon footprint reduction (Fig. 5.1).

An in-depth review reveals the specific areas of operations within an organization-wide system, including the pumps and processes in operations, equipment, fuel management, heating and cooling, power production or consumption, and renewables (Fig. 5.2).

Institutionalization of Resilient Energy Conservation in Operations

Municipalities embedding resilient energy strategies into the organizational culture derive multifold benefits. A range of staff from the city departments and beyond departmental silos are included in energy management, continually seeking energy-saving opportunities. The strategic resilience and sustainability plan promotes sustainable energy institutionalization. Each department uses the energy planning to organize activities and support budgetary choices based on the Quadruple Bottom Line elements. The plan holds the department leaders accountable for the continuous improvement of service delivery through the enforcement of the 5Is

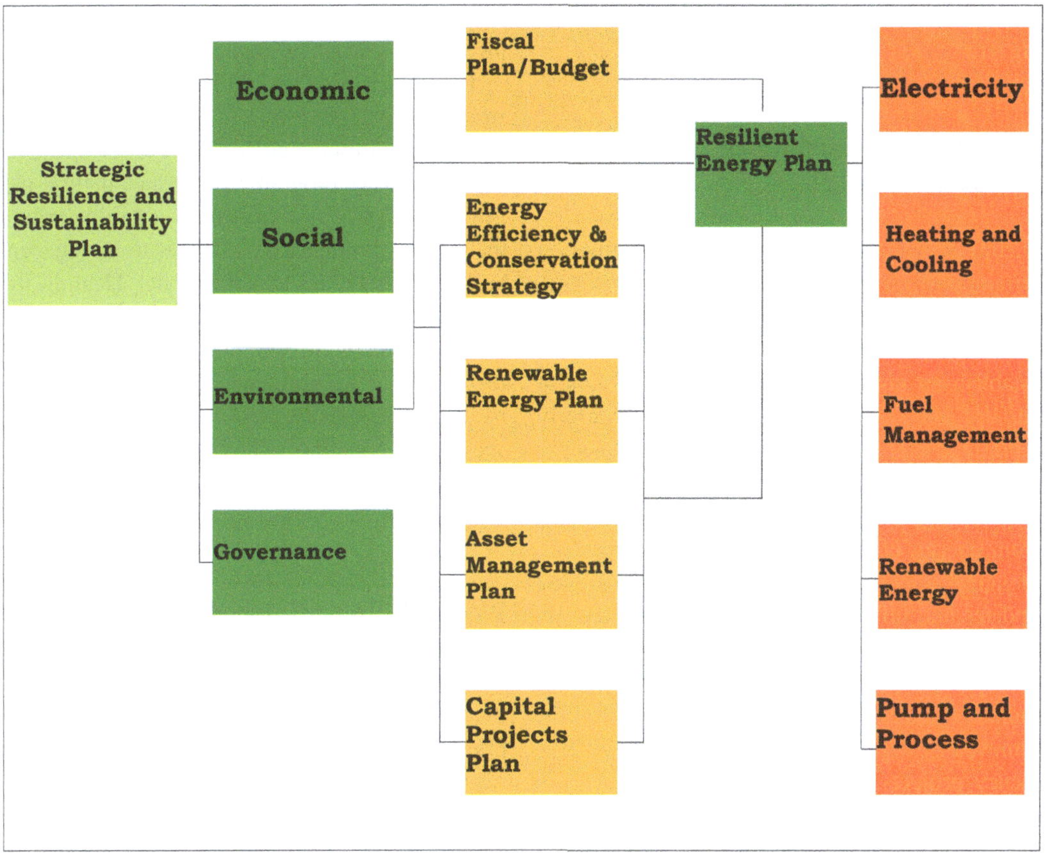

Fig. 5.1 Flowchart—resilient energy plan and resilience and sustainability planning and implementation

Electricity	Most local governments spend proportionally higher amounts annually on electricity than on fuels or natural gas. Addressing electricity consumption and cost related to the power supply is essential.
Heating and Cooling	Northern climate and climate change results in a number of days that require significant heating and cooling of buildings.
Fuel Management	Overall fuel consumption is measured in gallons of fuels or natural gas for CNG vehicles. Total costs continue to rise due to price increases.
Renewable Energy	Inherent to the resilience and sustainability approach is theadoption of renewable energy to reduce and eventually, discontinue the use of traditional sources of energy.
Pump and Process Equipment	Pumping operations can consume significant amounts of energy at wastewater and water treatment plants. It is the primary electrical demand in water plants and, in many cases, second only to aeration in wastewater plants.

Fig. 5.2 Table—key components of resilient energy systems

principles discussed previously in implementing strategies.

Measuring for Resilience and Sustainability

A commitment to reducing energy consumption is emphasized through tracking, measuring, observing, scrutinizing, and reporting progress, outcomes, and results. Municipal governments establish an inventory of electricity and natural gas use for all the city buildings and quarterly reporting. Employees identify opportunities for cost-saving measures after tracking, observing, scrutinizing, and reporting data and outcomes from energy efficiency and renewable projects. Organizational leaders emphasize results and updates and utilize savings for projects and service improvements. In addition to energy costs, administrators account for the carbon effects of municipal facilities and vehicle fleets. The CO_2 removed by organizations deploying energy efficiency projects contributes to the community's environmental, social, and governance benefits.

Empowering Staff and Encouraging Resilience in Innovation

Sustainable energy management enables staff to innovate in pursuit of energy management initiatives to reduce costs. Municipalities establish the sustainable energy team, consisting of staff from diverse departments involved in energy management decisions relative to energy use in facilities, fleets, or equipment. All types of energy consumption are scrutinized, including natural gas, steam, electricity, and fuel. The team incorporates a cross-section of personnel from diverse departments to develop a resilient and sustainable energy strategy, evolving as the future budgets warrant. The plans represent an opportunity to consume energy more efficiently, reduce greenhouse gas emissions, decrease energy and fuel use, lower energy costs, and support objectives of meeting the renewable energy target. It is advantageous for municipalities without their own power utility to partner with the private sector to meet the renewable energy target.

Employees use innovative, resilient energy strategies and initiatives to lessen the cost and improve environmental conditions. For instance, a heat recovery program may reduce natural gas usage by transferring heat from a facility where the process produces excess heat to an adjacent building during the winter months. Heating the adjoining building with sufficient waste heat to thoroughly heat the building without using natural gas may have an immediate return on investment. An organizational focus on resilient energy spurs innovation, novel designs, and produces an ongoing commitment from all city staff in every department.

Partnership for Resilient Energy Management

Prosperous programs involve innovative and pragmatic alliances to offset and leverage resources in the world fraught with financial instabilities. Financing initiatives require the engagement of the banking sector, the private sector, and the public sectors. One of the critical patterns in adopting resilience and sustainability is the advancement from planning to implementation. The partnership between municipalities and private sectors facilitates the implementation and execution of initiatives, policies, and programs. The municipal organizations and the community commit to the principles of resiliency and sustainability to support the natural environment, economic system, and social infrastructure. Steadfastness toward a resilient future is evidenced in collaboration and partnership throughout the community through sustainability programs, regional climate action plans, and other programs to share emerging practices, leverage resources, and positively transform regions and communities.

Municipal governments partner with power utilities through franchise agreements or procure energy from renewable sources. City administrators and managers use savings from reduced energy consumption to reinvest into renewable energy projects and purchases to meet the renewable energy target. A potential partnership exists in states with less restrictive energy markets, where a city may have a private firm or a utility to build and operate large-scale solar or wind projects. A potential large-scale wind project at the existing buildings, a solar project at an abandoned landfill site, a solar project at a property owned by cities, or a bio-digester project are all feasible projects in pursuance of renewable energy targets.

Resilient Communities and Organizations Spotlight: Michigan Cities' Resilient Energy Systems

Michigan was one of the states hardest hit by the economic recession from 2008 and the recent COVID-19 pandemic. Cities and counties handled the brunt of the financial disaster and the pandemic. Under unremitting economic crises, pandemic stressors, and climate change threats, administrators explore venues to reduce costs and adopt financial resilience strategies. Cities and counties seek to offset revenue losses by building a resilient and sustainable energy platform.

As part of strategies to increase resilience in communities, the pursuit of a robust energy program is a priority for organizations and communities. Energy and the associated costs are the common denominators. Comparable sustainable energy platforms exhibit similarities. However, there are also significant differences between these communities' approaches to resilient energy platforms. The cities of Ann Arbor, Dearborn, Holland, and Farmington Hills utilize diverse energy strategies to cut costs and increase renewable energy in their total energy use portfolio.

Common elements to these four cities' strategies are as follows:·

- Building awareness and education of the significance of energy consumption to the organization and community.
- Focusing on energy efficiency and energy conservation.
- Investing in renewable energy for production or acquisition of renewables in the energy portfolio.

Ultimately, the purpose of the resilience strategy is to provide sound policy guidance for future energy planning, proper asset management of energy infrastructure, buildings, and facilities, effective management of energy and transportation needs for the city. The following provides an overview of these elements for representative cities. Common factors pertain to other municipal resilient and sustainable energy initiatives around the United States and the world.

Resilient Energy in Ann Arbor, MI

Ann Arbor has a robust resilience and sustainability plan, concentrating on climate change and sustainable energy, in the areas of energy efficiency, energy management, community engagement, and renewable energy for the city's operation and the entire community. Resilience and sustainability objectives are institutionalized in the city's planning process and the progress updates in resilience and sustainability-related fields are regularly provided.

The resilient energy programs are both internal and external, supporting organizational operation efficiency efforts and community-wide energy efficiency, conservation, and renewable energy (City of Ann Arbor n.d.-a). Ann Arbor's energy programs include the following:

- Ann Arbor Renewable Energy Program, which features solar projects on fire stations, and solar markets, and exploratory plans of the project at the landfill site (City of Ann Arbor 2021).
- Ann Arbor Solarize, a program to incentivize community bulk buying of solar power (City of Ann Arbor n.d.-b).
- Ann Arbor has an active engagement in promoting electric vehicles and charging stations (City of Ann Arbor n.d.-c).

Through its outreach and education programs, Ann Arbor city employees promote energy efficiency and renewable energy to residents and business owners, with energy savings recommendations (City of Ann Arbor n.d.-b). The capacity of cities and counties to engage residents and business owners is beneficial in affecting energy consumption reduction and reducing the overall carbon footprint in the community.

Resilient Energy Strategies in Dearborn, MI

Other local governments in Michigan have also explored and implemented resilient and sustainable energy in their operations. The City of Dearborn (n.d.), with the population of almost 110,000 per United States Census Bureau (2021a) advances energy efficiency and conservation for all sectors, including municipal operations and private sector. The local government official in Dearborn invested in energy efficiency improvements, solar, and energy performance contracting for lighting in their firehouses.

The Dearborn representatives have been active in Michigan Green Communities Challenge, statewide renewable energy and environmental leadership initiatives launched in 2009, and where the City of Dearborn received a gold certificate designation for its efforts in sustainable energy and resilience (Dalbey 2016). David Norwood, the city's sustainability manager, noted the importance of sustainable energy initiatives in meeting the resilience and sustainability objectives "to reduce energy consumption and engage city staff in seeking innovative solutions for the city's operational efficiency and reduce the cost of service delivery for the community" (Norwood David. 2018. E-mail communication with author, January 22, 2018).

Resilient Energy Planning in Farmington Hills, MI

The City of Farmington Hills (2011, p. 1) leaders adopted Vision 2020, focusing on critical areas of interest to the community, including various sustainable energy management goals:

- Create a Community that Encourages and Embraces Energy Sustainability
- Advance Energy Efficient Practices
- Promote Alternative Energy and Fuels
- Improve Transportation Through Sustainable Practices

According to the most recent United States Census Bureau (2021b) data, over 80,000 people live in Farmington Hills. Farmington Hills administrators implemented an innovative performance contracting project for efficiency upgrades and conservation measures. Moreover, the city established the Commission for Energy and Environmental Sustainability (CEES) as an advising body to the City Council. Initially, the task of the commission was to advise the city on energy savings measures. The Commission expanded its scope of work to engage the community on energy and environmental issues (SustainableFH n.d.). Energy efficiency and energy management are central to the local government's resilient energy strategy. Concrete outcomes related to a successful deployment of resilient energy platforms positively affect communities and organizations.

Resilient Energy Planning in Holland, MI

City of Holland is a medium size community in Michigan with a population of over 33,000 (United States Census Bureau 2021c). The Holland Board of Public Works is a municipally-owned power utility in the city of Holland, Michigan (HBPW, 2017). Having a municipal utility presents a vital opportunity to advance the energy goals for the organization and community. In 2011, Holland developed a comprehensive community energy efficiency and conservation strategy, establishing a baseline and long-term energy scenarios (Garforth International et al. 2011). Holland Community Energy Plan promises to make the "community a national leader in energy security, affordability, sustainability and efficiency" (City of Holland n.d.-a, p. 1). The action items included in the strategy consist of a district heating program, Industrial Services—the "Holland Full Utility Service Bundle," a building energy labeling program, and community education and outreach—among others. These plans recognize all the community-wide and organizational outreach elements and a laser-like focus on energy efficiency and support for renewable energy.

One of the city's principal advantages over many other communities is owning a municipal power utility. The city pursued and funded energy projects through practical energy efficiency, including lighting upgrades and innovative downtown pedestrian lighting retrofits (the City of Holland n.d.-b). In 2012, the Sustainable Return on Investment (SROI) study examined "the affordability, reliability, social, economic, environmental, and health impacts of several generation options," with recommendation for the natural gas plant, supplemented with the "Power Purchase Agreements for renewable energy" (P21 2016b).

Furthermore, in 2014, the Holland City Council and the Holland Board of Public Works approved the sale of municipal revenue bonds to help fund building a combined-cycle natural gas power generating facility (HBPW 2014). The Holland Board of Public Works invested in renewables and energy efficiency projects, including downtown heating and cooling, and demonstrated the commitment to renewable energy through a 20-year contract with the Michigan Public Power Agency (MPPA) (HBPW n.d.). The HBPW built a new natural gas facility, with the excess thermal heat enabling expansion of the existing snowmelt system (P21 2016a). Immediately after its completion, the Holland Energy Park received "the Institute for Sustainable Infrastructure's (ISI) Envision Platinum award recognizing the sustainability of public infrastructure" (P21 2016c). Organizations achieve further savings with the use of grants, energy optimization rebates, and one-time investments.

Resilient Communities and Organizations Spotlight: Resilience in Storm-Water System, City of Philadelphia, PA

City of Philadelphia is a large city in Pennsylvania with a population of over 1,600,000 (United States Census Bureau 2021d). The City of Philadelphia Green City, Clean Waters program was established in 2011 to reduce stormwater runoffs, and combined sewer overflows using green infrastructure and systems (Philadelphia Water Department n.d.; Urban Land Institute 2018). Green and natural stormwater management systems deliver governance, economic, environmental, and social benefits and outcomes. It is less expensive to maintain green infrastructure in comparison to traditional gray infrastructure. The objective of green infrastructure investments is to create a paradigm shift in stormwater management to include proactive planning and an improved technical methodology of service delivery. The benefits of the resilient infrastructure to reducing stormwater runoff and combined sewer overflows are well documented, including reduced greenhouse gas emissions, job creation and private firms investment, and new funds invested in schools (Urban Land Institute 2018; Philadelphia Water Department n.d.).

Summary

Municipal elected and appointed officials pursue resilience of water, parks, parking, building, fleets, stormwater, utilities, energy, and other systems. Administrators continue to adopt new policies and strategies consistent with the Quadruple Bottom Line. Contemporary cities, townships, villages, and counties approve multi-year plans and strategies to implement comprehensive, robust, and measurable resilience and sustainability initiatives and projects. Implementing resilient projects reduces the cost and decreases carbon footprint and dependence on and consumption of fossil fuels.

Municipal organizations utilizing resilient and sustainable practices reduce the burden to taxpayers and create a more vibrant, healthy environment for residents with a strong economy and improvement to the governance. New technologies and cost-saving opportunities exist for municipalities to explore and utilize. Administrators, managers, and employees implement initiatives using the most effective long-term strategy to meet environmental, social, economic, and governance objectives for an organization and a community.

Energy is a substantial cost driver for organizations. Organizations strive to ensure savings in their operations by efficiently addressing energy costs. Notably, increased electricity costs are a budgetary burden for organizations. Additionally, decreased renewable energy costs enable a return on investment for renewable energy projects to be more economically and financially feasible. To achieve the most energy savings and integrate resilient and sustainable energy initiatives into greater resilience goals, local governments take a more systemic, holistic, and strategic Quadruple Bottom Line approach to energy planning and project implementation.

Organizations promote energy efficiency and renewable energy to bring awareness to the benefits of resilience and sustainable energy strategies. The energy strategies are developed on the principles of reducing electricity and natural gas consumption and associated costs, reducing transportation-related fuel consumption, and meeting renewable energy and greenhouse gas emissions targets. Lastly, cities explore water systems and stormwater management projects using green infrastructure to address environmental, health, social, and economic concerns from stormwater runoffs and combined sewer overflows.

Outcomes, Discussions, and Further Considerations

- Assess the implementation strategies for resilience and sustainability-related projects.
- Analyze the impact of resilient and sustainable energy strategies in organizations and communities.
- Consider the types of resilience and sustainability initiatives and projects.
- Evaluate the Quadruple Bottom Line strategies to project implementation and policy execution.
- Discuss the issues of water quality protection and stormwater management using green infrastructure.

References

Alexander G, Boyle G (2010) Introducing renewable energy. In: Boyle G (ed) Renewable energy: power for a sustainable future, 2nd edn. Indian Edition, Oxford

Alibašić H (2017) Measuring the sustainability impact in local governments using the quadruple bottom line. Int J Sustain Policy Pract 13(3):37–45

Alibašić H (2018) Sustainability and resilience planning for local governments: the quadruple bottom line strategy. Springer. https://www.springer.com/us/book/9783319725673

City of Ann Arbor (2021) Renewable energy. https://www.a2gov.org/departments/sustainability/Sustainability-Me/Families-Individuals/Pages/Renewable-Energy.aspx

City of Ann Arbor (n.d.-a) Ann Arbor solarize. https://www.a2gov.org/departments/sustainability/Sustainability-Me/Families-Individuals/Pages/Ann-Arbor-Solarize.aspx

City of Ann Arbor (n.d.-b) The green house: sustainability at home. https://www.a2gov.org/departments/sustainability/Sustainability-Me/Families-Individuals/Pages/The-Green-House-Sustainability-at-Home.aspx

City of Ann Arbor (n.d.-c) Electric vehicles. https://www.a2gov.org/departments/sustainability/Sustainability-Me/transportation/Pages/Electric-Vehicles.aspx

City of Dearborn (n.d.) Green dearborn. https://www.cityofdearborn.org/community/green-dearborn

City of Farmington Hills (2011) Vision 2020: a vision for a more sustainable and prosperous Farmington Hills. https://www.fhgov.com/Government/Reference/2020-Visioning-Reports/Energy-and-Transportation-Summary.aspx

City of Holland (n.d.-a) Holland community energy plan. https://www.cityofholland.com/DocumentCenter/View/466/Holland-Community-Energy-Plan-Brochure-PDF

City of Holland (n.d.-b) Downtown pedestrian lighting: LED retrofit project. http://www.cityofholland.com/sites/default/files/fileattachments/downtown_led_brochure_2011_0.pdf

Dalbey B (2016) Dearborn wins gold for energy, environmental initiative. Patch. https://patch.com/michigan/dearborn/dearborn-wins-gold-energy-environmental-initiative

Douthat T, Morgan JD, Alibašić H, Grant A (2020) Passive building characteristics, and summertime residential energy use: a spatial analysis of energy efficiency in Gainesville, FL. Build Environ 169:106542

Garforth International, LLC, City of Holland, Holland Board of Public Works (2011) Holland community energy efficiency and conservation strategy: creating a global competitive community. Project Work Team Report. http://www.cityofholland.com/sites/default/files/fileattachments/final_cep_for_suscom_sept_9_2011_for_website.pdf

Hardjono T, Van Marewijk MV, de Klein PD (2004) Introduction on the European Corporate Sustainability Framework (ECSF). J Bus Ethics 55(2):99–113. https://doi.org/10.1007/s10551-004-1894-x

Holland Board of Public Works (2017) About. https://www.hollandbpw.com/

Holland Board of Public Works (n.d.) Holland BPW approves 16.8 MW wind contract to diversify renewable energy portfolio. https://www.hollandbpw.com/about-us/news-releases/142-holland-bpw-approves-16-8-mw-wind-contract-to-diversify-renewable--energy-portfolio

Holland Board of Public Works (HBPW) (2014) Holland approves sale of bonds to finance new power plan. http://p21decision.com/2014/03/20/holland-approves-sale-of-bonds-to-finance-new-power-plant/

Kohl H (2000) Renewable energy sources on the rise. In: Wengenmayr R, Bührke T (eds) Renewable energy: sustainable energy concepts for the future. Wiley, Weinheim, pp 4–14

MacKay DJC (2009) Sustainable energy: without the hot air. UIT Cambridge, Cambridge

Ostrom E (1990) Governing the commons: the evolution of institutions for collective action. Cambridge University Press, New York

Philadelphia Water Department (n.d.) Green city, clean waters. https://water.phila.gov/green-city/

Power for the 21st Century. (P21) (2016a) P21: looking back. http://p21decision.com/2016/05/05/p21-looking-back/

Power for the 21st Century. (P21) (2016b) How the Energy Park will improve snowmelt. http://p21decision.com/2016/01/29/how-the-energy-park-will-improve-snowmelt/

Power for the 21st Century. (P21) (2016c) Holland Energy Park earns Envision® Platinum Award. http://p21decision.com/2016/07/20/holland-energy-park-earns-envision-platinum-award/

Sovacool BK, Watts C (2009) Going completely renewable: is it possible (let alone desirable)? Electr J 22(4):95–111. https://doi.org/10.1016/j.tej.2009.03.011

Stafford E, McCann RB (2020) How Utah cities are pursuing 100% renewable energy. GreenBiz. https://www.greenbiz.com/article/how-utah-cities-are-pursuing-100-renewable-energy

Stubbs W, Cocklin C (2008) Conceptualizing a "sustainability business model". Organ Environ 21(2):103–127. https://doi.org/10.1177/1086026608318042

Sustainable Farmington Hills (SustainableFH) (n.d.) About CEES. http://www.sustainablefh.com/Menu/About-CEES.aspx

UCLA Luskin Center for Innovation (2019) Progress toward 100% clean energy in cities and states across the U.S. https://innovation.luskin.ucla.edu/wp-content/uploads/2019/11/100-Clean-Energy-Progress-Report-UCLA-2.pdf

United States Census Bureau (2021a) QuickFacts Dearborn city, Michigan. https://www.census.gov/quickfacts/fact/table/dearborncitymichigan/PST045219

United States Census Bureau (2021b) QuickFacts Farmington Hill city, Michigan. https://www.census.gov/quickfacts/fact/table/farmingtonhillscitymichigan/PST045219

United States Census Bureau (2021c) QuickFacts Holland city, Michigan. https://www.census.gov/quickfacts/fact/table/hollandcitymichigan/PST045219

United States Census Bureau (2021d) QuickFacts Philadelphia city, Pennsylvania. https://www.census.gov/quickfacts/fact/table/philadelphiacitypennsylvania/PST045219

United States Conference of Mayors (USCM) (2017) 85th Annual meeting 2017 adopted resolutions. http://legacy.usmayors.org/resolutions/85th_Conference/proposedcommittee.asp?committee=Energy

United States Energy Information Administration (2019) U.S. renewable electricity generation has doubled since 2008. https://www.eia.gov/todayinenergy/detail.php?id=38752

United States Environmental Protection Agency (2019) Sources of greenhouse gas emissions. https://www.epa.gov/ghgemissions/sources-greenhouse-gas-emissions

United States Environmental Protection Agency (EPA) (2018) Green power partnership. Green power partnership top partner rankings. https://www.epa.gov/greenpower/green-power-partnership-top-partner-rankings

Urban Land Institute (2018) Green city, clean waters. Philadelphia, PA. https://developingresilience.uli.org/case/green-city-clean-waters/

Assessing the Intersection of Resilience and Sustainability

6

"Besides fuel, there is abundant material from which we might eventually derive power... But, whatever our resources of primary energy may be in the future, we must, to be rational obtain it without consumption of any material. Long ago, I came to this conclusion, and to arrive at this result only two ways, as before indicated, appears possible – either to turn to use the energy of the sun stored in the ambient medium, or to transmit, through the medium, the sun's energy to distant places from some locality-where it was obtainable without consumption of material... It is difficult to believe, but it is nevertheless, a fact, that since time immemorial man has had at his disposal a fairly good machine which has enabled him to utilize the energy of the ambient medium. This machine is the windmill. Contrary to popular belief, the power obtainable from wind is very considerable."

(Tesla 1900, p. 199).

Overview

The sixth chapter evaluates the effects of climate resilience, climate preparedness, climate mitigation, and climate adaptation planning processes and initiatives. Climate change generates further challenges for local government organizations demonstrated by extreme and unpredictable weather in recent years. While the impacts of climate change are regionspecific and are diverse in intensity and impact levels, they affect all aspects of communities. Chapter six extensively re-examines the intersection between resilience and sustainability in a theoretical and practical function. It describes the intersection between resilience and sustainability planning to advance positive outcomes.

Furthermore, it appraises strategies to integrate climate resilience and preparedness into short-term and long-term organizational and community-wide planning. Finally, the chapter offers an overview of the appropriate steps to integrate climate change strategies into emergency preparedness and disaster readiness. The ultimate goals of climate resilience planning are:

- The reduced operational costs.
- Decreased negative impact on the environment.
- Positive social effects on communities.
- Improved governance.

Keywords

Climate change · Climate resilience · Climate preparedness · Climate readiness · Climate adaptation · Climate mitigation · Emergency preparedness · Heatwave · Disaster mitigation Economic effects · Vulnerability · Global warming · Carbon footprint · Resilience Governance · Accountability

Key Questions

The goal of the sixth chapter of this book is to answer the following underlying assumptions and questions:

H. Alibašić, *Strategic Resilience and Sustainability Planning*, Sustainable Development Goals Series, https://doi.org/10.1007/978-3-030-91159-1_6

- What are the climate resilience, climate preparedness, climate mitigation, and climate adaptation strategies?
- How is organizational resilience planning related to sustainability strategic planning?
- What strategies do organizations utilize to integrate climate resilience into organizational planning, including short- and long-term objectives?
- How do municipal governments plan for climate change threats?
- What are the suitable steps to integrate climate change strategies into emergency management, readiness, disaster preparedness, and mitigation?
- How do communities evaluate and react to the impacts of climate change?
- What are the governance, economic, environmental, and social effects of climate change?
- How do communities increase resilience to threats of pandemics?
- How does the private sector respond to climate change threats?

Introduction

Administrators and managers face daily challenges as they manage the services delivery to residents and businesses in their communities. The new global and regional realities of climate change and extreme weather, coupled with the ongoing COVID-19 pandemic affect localities and organizations in their provision of an array of critical services, such as public safety, infrastructure, water, stormwater, energy supply, and waste management, to name a few. Municipal government administrators evaluate the vulnerabilities of communities to the climate change, the issue of greenhouse gas emissions and symptoms related to increased environmental pollution.

Climate change generates further challenges for local government officials exemplified through extreme and unpredictable weather patterns, including but not limited to heatwaves, intense rain events, more frequent flooding, changes in temperatures, snowstorms, more intense and less predictable hurricanes, fires, and droughts. Additional stressors occur due to the recent COVID-19 and shutdowns caused by the pandemic. While the impacts of the climate change are region-specific and are diverse in intensity and the impact levels, they affect all facets of healthy communities. Most notable impacts include water shortages, quality of water supply, and freshwater resources, power outages, and disruptions as demand for cooling increases during the heatwave events, demands on first responders, increased stressors on infrastructure, economy, service delivery, and vulnerable population.

Elected and appointed officials plan for and implement robust climate mitigation, adaptation, and climate preparedness strategies to ensure maximum community and organizational resilience. Cities and counties are vulnerable to climate change, and the implementation of climate adaptation and mitigation strategies is most effective on a localized scale in tandem with the regional strategy. Some of those measures may include:

- Developing a sustainable energy plan with energy conservation and efficiency strategies to reduce energy consumption and demand throughout the organization and plan for peak load demand in collaboration with power utilities.
- Setting a 100% renewable energy target for the city's operations by a target year. Diversified energy sources and decentralized power delivery are essential for local resilience and greenhouse gas emissions reduction.
- Reducing total fuel consumption in fleet and operations, and electrification of vehicles, coupled with more power charging stations.
- Establishing a target to increase the tree canopy cover and diversify the type of tree species planted to increase resilience to urban heat island effects and heatwaves.
- Effectively managing waste minimization, reduction, and recycling of materials.
- Providing exceptional public safety services and developing and implementing emergency and disaster preparedness plans and strategies, incorporating climate resilience vulnerability information, and pandemic preparedness plans.

- Reducing water consumption and protect water and other natural resources.
- Ensuring economic development, planning, traffic safety, lighting, streets, public safety, and engineering services are provided using a system-wide, holistic approach, in partnership with local and regional business and economic development agencies and other institutions for maximum outcomes.

As noted in the diagram, in recognizing the completeness and complexities of the systems, the elements of climate resilience and pandemic resilience are implemented through existing sustainability and climate action plans (Fig. 6.1).

It is compelling for organizations to adopt and undertake diverse methods to mitigate and adapt to the impacts of climate change. Climate change mitigation and adaptation strategies lead to a diversification of energy production and more resilient energy management, less centralized, systems, with fewer disruptions of power supply.

The initial step to climate resilience planning is the completion of the regional or local resilience report, with a specific and detailed understanding of climate change data, greenhouse gas emission, weather patterns, and localization of the climate change impact. For pandemic preparedness plans, organizations must take into consideration the significant level of disruption and expect coordination to occur between the

Fig. 6.1 Diagram—resilience and sustainability initiatives

national, state, and local levels of government and private health and public health institutions and businesses. Economic, social, environmental, and governance effects of disruptions are significant.

Defining Resilience to Climate Change, Pandemics, and Disasters

Resilience is characterized as the ability of organizations to prepare for the worst possible disaster scenarios, react to disasters and emergencies and recover by undertaking effective strategies to counter unforeseen events exacerbated by climate change, extreme weather events, and pandemics. The consequences of not planning and adequately preparing for potential disasters are devastating for the population, services, buildings, and infrastructure. Fiksel (2003) viewed system resilience through the lens of disruptions to the balance of systems. Fiscal constraints and the impact of climate change on the global economy, worldwide pandemics on cities, and the ability to deliver outcomes show significant disruptions.

Organizations deploy strategic resilience and sustainability planning to adapt, transform, and endure challenges, and continue the services provisions without interruptions. In a crisis, resilience and sustainability-related initiatives represent an opportunity to adjust and reevaluate the priorities and outcomes of the budget and fiscal plans. Rather than anticipating a national leadership on climate change, cities, villages, townships, counties, and other localities take proactive climate preparedness actions to pursue the interest of their constituents and residents. Cities have a unique role in providing services and deciding what policy options best fit the organizational and community framework. In addition to vulnerabilities to climate change and extreme weather events, cities are the most prominent contributors to carbon pollution. Fitzgerald et al. (2012) suggested the greenhouse gases gas emissions cause climate change and environmental challenges.

Cities are major contributors to greenhouse gas emissions, and the administrators strive to reduce the carbon impact on society. Byrne et al. (2006) argued that cities of industrial and more developed nations play an important role in addressing the negative consequences of pollution in urban industrial centers. The evidence of the impacts and consequences of climate change on the environment and societies are global in scope. The negative effects of COVID-19 have been widespread across the globe and most destructive in urban cities.

Climate Change Concerns for Organizations and Communities

An extensive body of scientific research points to the indisputable evidence of the harmful effects of industrial activities on climate, causing environmental and social disruptions. Extreme weather events, increases in global temperatures, sea-level rise, economic disruption, infrastructure damage, species extinction, and weather pattern changes are some of the ongoing and well-documented concerns and challenges for communities, population, and the society as a result of the changes in the climate (Blaikie et al. 1994/2014; Fletcher 2013; Hallegatte et al. 2011; IPCC 2013, 2014, 2018; Karl et al. 2009; Lindfield 2010; Mach et al. 2016; Malcolm et al. 2006; Mann et al. 2017b; Pecl et al. 2017; Segan et al. 2015; Stott et al. 2016; The World Bank 2012; Urban 2015; USGRCRP 2017; Visser et al. 2014). Dong et al. (2014) and Silva et al. (2013) examined the linkages between the human-caused carbon emissions, climate change effects, and mortality rate. Beyond the sea level rise and acidification of the oceans, the threats to Florida coastal and urban areas (Alibašić & Morgan 2020; Bloetscher et al. 2017; Chassignet et al. 2017; Climate Central and ICF International 2015) and Great Lakes (Environmental & Law Policy Center 2019; Kling et al. 2003) are studied at length and documented. The threats represented by climate change have significantly increased in recent years with impact on infrastructure, water, stormwater, energy supply, and entire systems (Bullock et al. 2014; Bloetscher 2012). The social and economic costs of climate change are tremendous and long-lasting (Yongyang and Lontzek 2019; Solomon et al. 2017; Solomon and Kopp 2018).

The most recent report of the Intergovernmental Panel On Climate Change (IPCC) from 2021 issued the starkest warnings of the irredeemable effects of climate change on the entire planet. The global community has been slow to react and to act decisively against the threats of climate change. The failure of national governments to address climate prompted leadership actions on a subnational level, mainly by local governments. With direct consequences forecasted, cities are at the forefront of those efforts to deal with climate change effusively.

Community-Wide Climate Resilience

Community-wide resilience preparedness considers emergency preparedness, infrastructure, transportation, energy planning, health, human resources, and public safety issues. The benefits of addressing and taking the climate change trends into consideration far outweigh the costs associated with climate resilience and preparedness planning. Furthermore, the resilience of the systems in crisis and during the disasters is the capacity, robustness, and adaptability to endure shocks and continue to thrive during and after the disasters. Hillman and Geunther (2021) expertly analyzed various literature on organizational resilience and Linnenluecke and Griffiths (2012) discussed organizational resilience to climate change. Conversely, Alibašić (2018a, b, c) discussed the importance of climate change planning for communities in connection to resilience and sustainability strategic plan to include all the elements of climate change analysis, vulnerability study, and impact assessment. Moreover, Zautra et al. (2008) offered an evaluation of the integrative approach to community resilience.

Weather patterns can no longer be predicated upon existing models and the impacts on regions, cities, and especially urban areas are immense. Besides, the more frequent severe weather events and changes in precipitation and temperature patterns impede the social system, governing, ecosystem, and the economy. As a direct result of changing climate, there are more frequent and severe heat waves, hurricanes, tornadoes, fires, excessive rain events, and flooding, and changes in temperature and precipitation pattern impending social systems, ecosystems, and the economy with severe negative consequences on urban areas (Alibašić 2018b, c). Moreover, the major climate trends such as air and water temperature variations and increases, droughts and dry seasons, the frequency and intensity of storms, and floods impact a spectrum of sectors. The principal feature of climate change actions is the ability of organizations to prepare all systems to be more agile and adaptive to extreme weather events and disasters.

Climate Change Resilience and Emergency Preparedness

A combination of climate change and pandemic threats represent a whole set of extra challenges in emergency planning, preparedness, and disaster mitigation for municipalities. Studies by Gimenez et al. (2017), Son et al. (2017), Tveiten et al. (2012) analyzing the connection and relevance of resilience in emergency management offered valuable insight into the dynamic of emergency preparedness and disaster planning.

At a minimum, emergency plans incorporate the latest sets of data on climate vulnerabilities, greenhouse gas emissions, and costs to understand impacts and offer solutions and alternative approaches. A changing climate generates many challenges for state, local, and tribal governments as elected leaders, planners, and resource managers considering mechanisms for ensuring community resilience and preparedness.

In response to heatwave events, floods, pandemics, and fires, cities and counties adjust their emergency action guidelines to coordinate services with the nonprofit agencies and utilities,

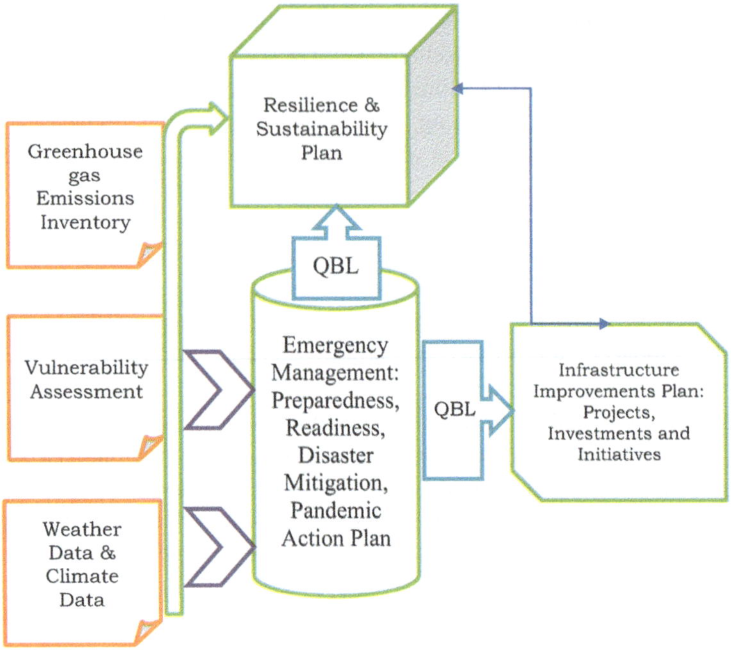

Fig. 6.2 Combining resilience and sustainability planning with emergency management and disaster preparedness

public health institutions and agencies, hospitals, facilities, and cooling centers. Local governments respond holistically to heatwaves and hazardous rain events and have more control over the events arising from climate change and extreme weather events by connecting the resilience and sustainability plan directly to emergency management, disaster planning, and mitigation. Having the accurate climate data, vulnerability assessment, information about the history of weather patterns and events and infrastructure plans are critical components of resilience and sustainability planning as shown in Fig. 6.2. Each element has to be carefully incorporated into the resilience and sustainability strategic planning with governance, economic, environmental, and social factors considered.

The Pandemic Resilience Planning

The recent global events caused by the COVID-19 pandemic put a special spotlight on cities and urban centers. Due to the density of buildings, an enormous number of employees working in cities, massive transportation and mobility issues with moving millions of people daily, COVID-19 placed an extraordinary strain on human resources and infrastructure. Crisis management under COVID-19 brings an extraordinary set of challenges not faced before (Van den Oord et al. 2020). Noting the economic and social changes that may potentially occur in the post-pandemic world, Välikangas and Li (2020) posited the question of how the pandemic may advance resilience. Furthermore, Dzigbede, Gehl and Willoughby (2020) evaluated how the local governments can strengthen their disaster resilience and responses to the COVID-19 pandemic. The COVID-19 pandemic will continue to represent an ongoing threat due to the politicization of responses around the United States and globally, undermining mask policies, responses, and mass vaccination efforts. The role of municipalities in such a dynamic position where they must rely on national and state coordination is very precarious, and the study of the local governments' handling of pandemics and emergency management and preparedness deserves a separate and more in-depth analysis and treatment.

National and Subnational Climate Resilience Policies and Initiatives

A lack of national resilience policies and programs enable decentralized approaches to climate resilience planning, bringing to light the relevance of local governments and their policies toward sustainability, resilience, and climate preparedness. In discussing scaling climate adaptation strategies in urban settings, Brugmann (2012) viewed resilience as the ability of organizations to perform predictably. In examining several municipalities in Canada, Burch (2010) noted varying levels of success in carbon footprint reduction in these communities.

In the United States, during President Obama's administration, there were more aggressive national-level efforts to address climate change. Earlier, the US Conference of Mayors established the Climate Protection Agreement with an overarching goal to encourage cities to deploy climate mitigation strategies and reduce the organizational carbon footprint of city organizations with the commitment from over one thousand US mayors, leading to more participation by the local governments (USCM 2005). Under President Obama, local, tribal, and state leaders had a prominent role in planning for climate preparedness, working directly with the White House staff.

In 2013, by Executive Order 136531, Preparing the United States for the Impacts of Climate Change, President Obama established the State, Local, and Tribal Leaders Task Force on Climate Preparedness and Resilience (Task Force 2014). The task force members were asked to examine all venues of responses and preparedness to address climate change trends and threats and recommend the climate strategies for the federal government to better support local and state actions to make communities more resilient. Additionally, task force members reached out to thousands of government organizations, universities, and other stakeholders, seeking recommendations, focusing on scientific and policy approaches to climate preparedness and resilience strategies (Task Force 2014). In 2017, the United States withdrew from Paris Climate Accord, and other environmental protection and climate change initiatives were undermined by the administration (Davenport 2017; United States Department 2017).

City and county administrators continued to pursue climate preparedness and resilience strategies. As observed by Alibašić (2018a) and Gordon (2016), cities have undertaken countless actions in addressing climate change-related threats. The scaling of the climate resilience efforts to cities and counties levels is emphasized in outcomes from climate change-resilient activities. Cities' initiatives are designed to reduce greenhouse gas emissions, mitigate the impact of climate change, or adapt to new climate realities. Additionally, 364 US mayors agreed to cut greenhouse gas emissions per Paris Climate Accord requirements (Climate Mayors 2017). Furthermore, over 250 cities adopted the 100% renewable energy targets, and the EPA's top 30 lists are regularly updated, presenting renewable energy commitments from public and private organizations (USCM 2017; US EPA 2021a, b, c).

According to the United States Environmental Action Protection Agency (US EPA)(2021) report from July 26, 2021, "the combined annual green power use of EPA's Top 30 Local Governments amount to nearly 5 billion kilowatt-hours of green power, which is equivalent to the annual electricity use of more than 464,000 average American homes" (para. 1). As of July 1, 2019, there were an estimated 140 million housing units in the United States (United States Census Bureau 2019). Clearly, the public and the private sectors are still far from reaching the 100% renewable energy goals and the federal government's involvement is instrumental. Under President Biden, the United States reaffirmed its commitment to the goals set under Paris Climate Accord, setting 2030 greenhouse gas pollution reduction targets, and attempting to introduce climate change risk assessments for businesses, banks, and financial institutions (The White House 2021; U.S. Securities and Exchange Commission 2021; Brunetti et al. 2021).

The 100% renewable energy targets are intended to enable cities to mitigate and reduce impacts from energy consumption in city operations to green energy demand. These efforts within and outside organizations engaged in climate resilience and preparedness are necessary to prepare for inescapable disasters. Constituents at a local level, cities, and communities expect their appointed and elected officials to meet challenges and objectives to address concerns and prepare for the climate change-related impacts.

Cities and Counties Government Roles Reinvented

As cities take a more active role in climate preparedness and resilience, the local zoning and land-use policies are reinvented; new approaches to the procurement of energy and power supply and production are adopted (Alibašić 2020a, b). By taking a more proactive role and in some cases partnering with the private sector or academic institutions, local government units are less dependent on federal and state policies.

The reinvented role of local governments and the interaction on a subnational, national, and international scales are critical to creating synergies in concerted attempts to reduce, minimize, and ultimately remove the adverse impacts of climate change. The collective body of local government policies, projects, and programs leaves an impressive impact on the environment, society, economy, and good governance. Local leaders and administrators draw from the shared experience of cities worldwide, leading to better governance of local resources. Environmental, economic, social, and governance concerns are no longer viewed as localized issues. Cities continue to pursue resilience policies by combining sustainability-related efforts, countering adverse climate change effects, increasing resilience and resistance, and deploying adaptation strategies.

The local governments' cutting-edge planning for climate resilience in the United States is evident in cities including Ann Arbor, MI; Austin, TX; Baltimore, MD; Boston, MA; Chicago, IL; Dubuque, IA; Eugene and Portland, OR; city and county of Los Angeles and San Francisco in California; Philadelphia, PA; New Orleans, LA; Broward, Miami-Dade, Monroe, Orange, and Palm Beach Counties and cities of Orlando, Pensacola, Sarasota, and Tallahassee in FL, and New York City, NY, to name a few (City of Ann Arbor 2020a, b; City of Austin 2014, 2015, 2018; Baltimore Office of Sustainability n.d.; Broward County 2015; City of Boston 2014, 2017; City of Chicago n.d., 2008, 2019; City of Dubuque 2013, 2020; City of Eugene n.d., 2010, 2020; City of Los Angeles 2015a, b, 2018; County of Los Angeles 2015, 2021; Miami-Dade County 2021; City of New Orleans 2017; City of New York 2017, 2021; City of Orlando 2018; City of Philadelphia 2016, 2017; City of Portland and Multnomah County 2017; City of Pensacola 2018a, b; City and County of San Francisco 2013, 2017, 2021; Southeast Florida Regional Climate Change Compact Counties 2012, 2021; City of Tallahassee 2019). Other cities and counties in the United States and Canada, and around the world are making significant inroads and progress toward climate resilience planning and implementation of the resilience initiatives. For instance, the City of Pensacola (n.d.) task force on climate adaptation and mitigation recommended a set of climate resilience strategies to the council members and the Mayor. In Arkansas, the City of Fayetteville elected officials published a document entitled Arkansans Can Take Steps to Respond to Climate Change is a call to the state residents to be proactive in combating the threats of a changing climate (City of Fayetteville n.d.). The table lists some of most operational climate resilience plans, climate preparedness, and climate mitigation and adaptation action strategies, with a sample of related programs and departments overseeing them (Table 6.1).

Resilient Communities and Organizations Spotlight: City of Pensacola, Florida Climate Resilience and Sustainability Planning

The city of Pensacola in Florida is a medium-sized community with a population of over 54,000 (United States Census Bureau 2021). The City of Pensacola's (n.d.) Council appointed the Task Force on Climate Adaptation and Mitigation

Table 6.1 A sample of local governments with climate resilience plans, strategies and initiatives

Local Government, State	Year of release	Title of the plan	Department/position	Programs
City of Austin, Texas	2015 & 2018	Austin Community Climate Plan & Climate Resilience Action Plan for City Assets and Operations	Office of Sustainability	Climate Change, Sustainability & Utility Infrastructure, Transportation, Community Facilities
City of Ann Arbor, Michigan	2020	A2Zero Ann Arbor's Living neutrality Carbon Plan	Sustainability and Innovation	Renewable Energy, Switching Appliances and Vehicles to Electric, Energy Efficiency, Mobility, Transportation, Use, Reuse and Disposal of Material, People and Place Resilience, Equity, Carbon Offset, Internal Carbon Price
City of Baltimore, Maryland	2013	Baltimore Climate Action Plan	Office of Sustainability	Climate Action, Sustainability
City of Boston, Massachusetts	2014 2017	Greenovate Boston 2014 Climate Action Plan Update & Resilient Boston	Chief Resilience Officer	Resilience & Equity/Racial Equity, Climate Change, Environment, Governance, Adaptation, Connectivity
Broward County, Florida	2015 & 2020	Climate Action Plan: Local Strategy to Address Climate Change & 2020 Broward County Climate Action Plan	Chief Resilience Officer/Environmental Protection and Growth Management Department	Policy, Healthy County, Transportation, Built Environment, Energy Resources, Natural Systems, and Water Resources
City of Chicago, Illinois	2008, n.d. & 2019	Climate Action Plan: Our city, out future Resilient Chicago: A Plan for Inclusive Growth and Connected City	Department of Planning and Development/ Sustainable Development Division/Sustain Chicago	Environment and Sustainability; Strong Neighborhoods; Robust Infrastructure; Prepared Communities
City of Detroit, Michigan	2020	Detroit Climate Strategy	Office of Sustainability	Energy, Water, Transportation, Air Quality, Parks, Public Engagement
City of Dubuque, Iowa	2013 & 2020	Dubuque Community Climate Action & Resiliency Plan 2013 & City of Dubuque Climate Action Plan 2020	Sustainable Dubuque	Building & Energy; Transportation & Land Use; Solid Waste; Water, Wastewater & Flooding; Climate Health & Safety; Food; Greenspace & Tree Canopy; Climate Economy; Climate Action Capacity
City of Eugene, Oregon	2010 & 2020	A Community Climate and Energy Action Plan for Eugene & Climate Action 2.0	Office of Sustainability	A Data-Driven Roadmap for Climate Change; Local Impacts of Climate Change; Pathway to the CRO; Commitment to Action; Reducing Local Emissions; Eugene's Consumption Emissions; Climate Resiliency; Community Capacity; Equity Panel; Eugene Climate Collaborative; Mayor's Climate Recovery Ordinance Ad Hoc Work Group; Individual Actions; Community Engagement

(continued)

Table 6.1 (continued)

Local Government, State	Year of release	Title of the plan	Department/position	Programs
City of Los Angeles, California	2015 & 2018	The pLAn & Los Angeles Climate Action Report: Updated 1990 Baseline and 2013 Emissions Inventory Summary & Resilient Los Angeles	Mayor's Office of Resilience	Safe and Thriving Los Angelenos; Strong and Connected Neighborhood; Prepared and Responsive City; Pioneering and Collaborative Partner; Sustainability, Resiliency & Preparedness
Los Angeles County, California	2015 & 2021	Final Unincorporated Los Angeles County Community Climate Action Plan (CCAP) 2020 & COVID-19: Economic Resiliency & Los Angeles County Community Disaster Resilience	Department of Regional Planning	Planning, Climate Action; Community Disaster Resilience
Miami-Dade County, Florida	2021	Strengthening Resilience in Miami-Dade County	Office of Resilience/ Chief Resilience Officer	Sea-level rise & Flooding; Building Efficiency; Greenhouse Gas Inventory; Sustainable Buildings Program
City of New Orleans, Louisiana	2017	Climate action for a Resilient New Orleans	Mayor's Office of Resilience and Sustainability	Resilience and Sustainability: Greenhouse Gas Inventory and Reduction Strategies; Energy Use Modernization; Improve Transportation Choices; Reduce Waste; Culture of Awareness; Action Implementation and Monitoring; Inventory Methodology
City of New York, New York	2017 & 2021	1.5 °C: Aligning New York City with the Paris Climate Agreement & State of Climate Knowledge & Stormwater Resiliency Plan & Climate Resiliency Guidelines	Mayor's Office of Climate Resiliency	Climate & Energy Coastal Resiliency; Stormwater Plan; Climate Resiliency
City of Pensacola, Florida	2018	Climate Action Recommendations A Blueprint for Addressing Climate Change at the Municipal Level	Sustainability Coordinator & Climate Adaptation and Mitigation Task Force	Transportation; Built Environment; Greenhouse Gas Emissions Reduction Goals; Emergency Planning; Local Utilities; Economic Development and Resilience; Public Health; Outreach
City of Philadelphia, Pennsylvania	2016 & 2021	Growing stronger: Toward a climate-ready Philadelphia & Philadelphia Climate Action Playbook	Office of Sustainability	Energy Benchmarking, Climate Adaptation Planning; Carbon pollution reduction; Utilizing nature as solution; Adapting to changing climate
City of Portland, Oregon	2015	Climate Action Plan: Local Strategies to Address Climate Change (Portland and Multnomah County 2015 Climate Action Plan)	Portland Bureau of Planning and Sustainability	Planning, Sustainability, Climate Action

(continued)

Table 6.1 (continued)

Local Government, State	Year of release	Title of the plan	Department/position	Programs
Orange County, Florida	2021	Sustainable Operations and Resilience Action Plan	County Administration/Chief Sustainability and Resilience Office	Energy & Climate Action; Buildings & Infrastructure; Water Use & Quality; Mobility & Fleet; Supply Chain & materials Management; Trees & Land
City of Orlando, Florida	2018	Community Action Plan	Office of Sustainability & Resilience	Clean Energy, Green Buildings, Local Food, Livability, Solid Waste, Transportation, Water
City and County of San Francisco, California	2013 & 2021	San Francisco Climate Action Strategy. Department of Environment: 2013 Update 2021 Documents	Department of Environment	Climate, Energy, Transportation, Zerto Waste, Urban Forest and Greening; Sustainable Neighborhood; Sea-Level Rise; Climate Resilience
City of Tallahassee, Florida	2019	Tallahassee Community Resilience Plan	Sustainability and Resilience Division	Public Safety and Preparedness; Hazard Mitigation and Climate Adaptation; Equity and Social Cohesion; Planning and Integration
Southeast Florida Regional Climate Change Compact Counties, Florida	2012 & 2021	Southeast Florida Regional Climate Change Compact Counties—Regional Climate Action Plan	Southeast Florida Regional Climate Change Compact is a regional collaborative including Broward, Miami-Dade, Monroe and Palm Beach Counties	Climate Resiliency; Public Engagement

to recommend climate resilience strategies to the council members. Risks from climate change to Florida's coastal towns and cities and elsewhere are well-documented in the form of heatwave events, flooding, public health, extreme weather, and more intense and frequent hurricanes (Brady 2018; Luber et al. 2014; US Global Change Research Program 2017; Walsh et al. 2014).

The Task Force finished the report for the city and included sets of recommendations to create a more resilient community in public and private sectors, including transportation, water resources, utilities, natural resources, agriculture, and emergency management (City of Pensacola, 2018a, b; Megginson 2018; Morrison 2019, 2020). Also, the recommendations asked the city leadership to commit to renewables, reduce greenhouse gas emissions and warned of rising sea levels and extreme weather to infrastructure and vulnerable populations, suggesting adopting more stringent planning to deal with these threats. After the City Council adopted the recommendations, the city implemented several initiatives, including joining the International Council for Local Environmental Initiatives (ICLEI)—Local Governments for Sustainability, setting a 30% renewable energy target by 2035, creating a sustainability office, and hiring a sustainability coordinator.

The City of Pensacola completed a sea-level rise vulnerability assessment using the following ten criteria: Assessment Criteria: Hydric Environment, Critical Infrastructure, Parcel Level Land Use, Emergency Management, Potable Water, Sanitation and Sewer, Stormwater, Transportation, FEMA Floodplain, and Social Vulnerability (Clearview Geographic and Deady 2021). The assessment evaluates risks and vulnerabilities along the coastal areas and shows potential sea-level rise impact using interactive map tool.

Resilient Communities and Organizations Spotlight: Florida Counties Embedding Climate Resilience

A vital component of a successful implementation of resilience and sustainability strategic planning is organizational commitment. The Broward County (2015), with a population of almost 1,950,000, according to the United States Census Bureau (2021), adopted its first Climate Action Plan in 2015, focusing on resilience and sustainability actions. The Broward County Government's (2020) Climate Action Plan is divided into seven distinct elements: Policy, Healthy County, Transportation, Built Environment, Energy Resources, Natural Systems, and Water Resources. Resilience is managed by Environmental Protection and Growth Management Department and Resilience is divided into Water Policy and Planning, Climate Policy and Planning, and Science and Data (Broward County Government 2021a, b). The division oversees climate, energy, and sustainability programs (Broward County Government 2021c).

The commitment in staffing and resources exists to plan and implement resilience and sustainability strategies and initiatives. Broward County's budgetary, staffing, and organizational commitment to strategic resilience and sustainability planning are critical components to successful resilience and sustainability policies implementation.

Orange County (2021) developed the sustainable operations and resilience action plan during the COVID-19 pandemic with a range of support, involvement, and input from diverse stakeholders. The plan is divided into several sections: Energy & Climate Action; Buildings & Infrastructure; Water Use & Quality; Mobility & Fleet; Supply Chain & Materials Management (Orange County 2021). Earlier, Escambia County (2017) in Northwest Florida published an extensive, county-wide resilience adaptation plan. The plan relies on vulnerability and sea-level rise risk assessments and analysis. With most of its 160,000 residents residing near the shoreline and coastal areas on the east-central coast of Florida, Martin County (2021) published a sea-level rise risk analysis detailing climate change risks and extensive impact assessments by sectors.

Resilient Communities and Organizations Spotlight: Michigan Cities Climate Preparedness and Planning

In contrast to Florida's cities and counties' climate action planning, there are similarities and differences to Michigan cities' and counties' approaches to climate resilience. Communities of Ann Arbor, Detroit, Grand Haven, and Traverse City in Michigan, among others, have noteworthy climate resilience initiatives. To illustrate, the City of Detroit's climate strategy encompasses all the aspects of sustainability and climate resilience planning, including climate adaptation and mitigation strategies (City of Detroit 2021). To demonstrate, climate strategy includes plans to make energy and water more affordable, improve access to dependable transportation, make parks and green spaces welcoming, improve air quality and reduce cases of asthma, build resilience into daily lives to minimize climate impacts from flooding, heat waves, and extreme weather, and have inclusive public engagement to ensure equity integration into municipal and community climate action (City of Detroit 2021).

The City of Ann Arbor's (2020a, b) A2Zero Carbon Neutrality plan has six distinct strategies, including renewable energy target, switching appliances and vehicles to electricity, improving energy efficiency, increasing mobility, improving transportation, better use, reuse, and disposal of materials, increasing resilience of people and community, more equitable community, carbon offsetting, and setting an internal carbon price. Moreover, the city of Ann Arbor has significant funding committed to staffing and supporting sustainability and climate action planning. The Mayor proposed a tax increase through Community Climate Action Millage to assist the city's transition to a carbon-neutral future and to meet the community-wide and operational renewable energy target (Fleming 2021).

While approaches to climate resilience planning by these two cities appear divergent, the end goals and results are similar as both cities focus on climate change using both climate mitigation and adaptation strategies by embedding climate resilience in their plans. Climate resilience and sustainability strategies are inseparable as the cities attempt to address both climate adaptation and climate mitigation.

The Traverse City developed a climate action plan with SEEDS, Inc. with funding from the Department of Energy (City of Traverse City 2011). The program includes Ten Strategies for Action as guidelines to meet the climate and energy goals. In Grand Haven, local planners, and administrators have developed a comprehensive Resilient Master Plan. The Master Plan, among other issues, covers climate change threats to the Great Lakes, using scientific research to support findings and recommendations. Some of the noted trends and risks are increased precipitation and storminess, a variability of lake water levels, and water temperature. As a city on the Great Lakes shoreline, the administrators took a long-term view on planning mechanisms, addressing threats and obstacles in a comprehensive, dynamic, and holistic approach. Embedding climate resilience strategies into the Master Plan is a practical resilient planning approach.

Moreover, in addition to featuring Grand Haven, Resilient Michigan organization lists other communities in Michigan with resilient master plans in place, including cities of Holland, Beaver Island, Bridgman, East Jordan, Ludington, Macomb & St. Clair, and St. Joseph (Resilient Michigan, n.d.; Resilient Macomb 2016). To illustrate, Resilient Monroe Resource Atlas is a land-use master planning and community design project for the City of Monroe, Frenchtown Charter Township, and Monroe Charter Township. Among other tools and analyses, it includes a review of the urban heat island effect, heat sensitivity and exposure assessment, flood vulnerability assessment, and drainage stormwater management (Resilient Monroe, 2013). The report's authors wrote that climate scientists noted "that the Monroe Community and Southeast Michigan can expect more frequent storms of increasing severity in the decades ahead. The total amount of rainfall is also likely to increase. However, climate models suggest that the precipitation will be more concentrated in the winter, spring, and fall seasons as well as localized intense storms at almost any time" (Resilient Monroe, 2013, pp. 8–10).

In recent years, both the planning and investments in infrastructure made by cities were tested with the flood events and the extreme heatwaves in the summers. Communities have made preemptive investments in resilience, sustainability, and emergency planning to avoid further costly damages to the infrastructure and resources and implementing a variety of climate mitigation and adaptation strategies, including but not limited to:

- Developing energy conservation and efficiency strategies to reduce energy consumption and demand throughout the organization.
- Investing in production from renewable sources to diversify energy sources, as an essential step toward local resilience and for greenhouse gas reduction.
- Reducing total fuel consumption.
- Setting a goal to increase its tree canopy cover and diversify the type of tree species planted.

Local governments have been partnering with local nonprofits, grass-root organizations, residents, utilities, and academic institutions and developing climate change assessments, data reports, and resilience plans to further concentrate on climate, energy issues, economy, transportation, and infrastructure and to inform decision-makers in the areas of sustainability, ordinances, policies, and adaptation and mitigation strategies. The resilience and sustainability plans acknowledge climate change and serve to prepare the community to be more robust, agile, and adaptive to extreme events and disasters.

Private Sector Perspective: Climate Resilience and Sustainability Planning

Addressing climate change is impossible without the participation and active role of the private sector. For some time, companies, businesses, entrepreneurs, and corporations recognized the threats of climate change and attempted to counter the negative effects of climate change (Agrawala et al. 2011; Alibašić 2018d; McGregor 2017). The private sector leaders, CEOs, and managers recognize the negative impact in contributing to the overall climate change from greenhouse emissions associated with production and service delivery. Griffin (2017) and Riley (2017) reported that 100 corporations are responsible for 71% of global emissions and nearly one trillion tons of greenhouse gas emissions. Major contributors to climate change and negative environmental impacts are corporations, advanced by globalization (O'Brien and Leichenko 2000). One of the proactive approaches by corporations is investments in renewable energy. The US EPA (2021b) Green Power Partnership Fortune 500 Partners List for renewable energy shows that as of July 26, 2021, "more than 50 billion kilowatt-hours (kWh) of green power," which is "equivalent to the annual electricity use of more than 4.7 million average American homes" is purchased and used by major corporations. The list of 71 corporations includes Google LLC, Microsoft Corporation, Intel Corporation, Apple Inc., Walmart Inc., Bank of America, Wells Fargo, General Motors LLC, and others, with some of them on the 100% Renewable Energy partners list (US EPA 2021b, c).

Climate Resilience and Economic Development

There are positive outcomes from linking climate resilience planning, sustainability, and organizational efficiency to promote resilience and policies to decrease carbon emissions, lower costs and improve economic development and growth opportunities. Fitzgerald (2010) noted the linkages between sustainability and climate change initiatives. Importantly, it is incumbent for local governments to recognize the realities of climate change in resilience reports and plans. As a way of illustration, local governments link climate change concerns in the resilience and sustainability plan, supported by the realities of impact on jobs, business, and economic development, essential to resilience planning. Climate strategies and plans connecting the climate change effects to the number of potential jobs consider the full effects of greenhouse gas emissions and losses to farming, mining, tourism, energy production, and other sectors. City of Dubuque's (2020) climate action plan has a section dedicated to climate economy, calling to attention the inexplicable linkage between economy and climate change and stating that reducing greenhouse gas emissions can lead to increased economic growth and development. According to the plan, "between 2003 and 2018, Dubuque was able to decrease its GHG emissions by 27% while growing its economy by 78% and employment by 12%" (City of Dubuque 2020, pp. 11–13). The plan emphasizes the importance of equity in economic development and how climate change affects low-income individuals in communities, bearing "a greatly disproportionate share of the costs" (City of Dubuque 2020, pp. 11–13).

Embracing and Adapting New Systems and Technologies

The importance of adopting new technology and system improvement is exemplified through local governments' ability to provide services in times of crisis. The COVID-19 crisis has upended the entire spectrum of operations, systems, economies, societies, and governance. Pandemic has affected people, individuals, and human resources all over the world. The local government organizations regularly evaluate and adapt to new systems and embrace technological advancements and new technologies as operations moved to online services. Cities unwilling to adapt and embrace new technologies face diminishing returns on investments and losses. Adaptative systems in complex, dynamic situations and changing circumstances are capable of delivering results and meeting priorities. Greater collaboration in the technological systems delivers synergies and improvement to organizations, from sharing knowledge, training opportunities, engineering practices, and cost-savings from measures to increase resilience and improve sustainability. Pandemic and climate change threats challenge existing paradigms on how organizations adapt and evolve operating. Throughout the COVID-19 pandemic, many cities moved some of their services online, while most school districts, colleges, and universities adapted to offering online and distance learning with some degree of success.

Resilience Through Transformation and Awareness

There are compelling reasons for local units of government and other organizations to embrace and benefit from climate resilience planning from sustainable energy outcomes, cultural transformation, and positive changes in operations and service delivery. There is an overall belief that strategic resilience and sustainability planning positively impacts operations and meets community expectations for change. Policymakers and administrators embrace ambiguities in adopting transformational changes and strategies in organizations. Organizations undertake transformative measures to build a stronger and more resilient community ready to respond to changing demands and surrounding economic and environmental threats and uncertainty. The transformation toward resilience and sustainability strategic planning is an opportunity to build more resilient and sustainable organizations and communities.

Summary

The focus of this chapter was the interconnection between climate resilience and sustainability in planning, initiatives, programs, and strategies in local government organizations and communities. If properly utilized, tracked, measured, and compared to actual budgetary results and fiscal performance, resilience and sustainability strategies produce tangible, long-term positive effects and improve the overall effectiveness of service delivery. The ultimate goals of resilience planning are improving governance, reducing the cost of operations, decreasing the environmental impact, and positive social effects on communities. Analyzing the effectiveness of climate preparedness and its implications on the social, economic, ecological, and governance of organizations leads to a better perception of and confidence in organizational and community strategic resilience planning. An additional layer of complexity is evident with the ongoing threats of the COVID-19 pandemic.

Resilience planning at a regional level allows for outcome-driven partnership and sharing of responsibilities and resources. By using a dynamic approach to resilience planning, organizations steadily adapt to shifting governance, economic, environmental, and social conditions. Resilient organizations and communities continually build upon existing plans, layering and preparing to adapt and mitigate. Accordingly, public service practitioners examine the current policies to identify strategies and targets to meet climate resilience outcomes. Embedding climate resilience strategies into existing plans or updated plans is an efficient way of committing to the climate resilience action, as long as organizations connect the implementation and projects to their annual budget process. Climate preparedness, readiness, and resilience planning are at the forefront of local government's actions to create a more resilient future for communities. Resilience is observed and implemented contextually as an instrument leading to improved governance of environmental, social, and economic resources. Moreover, the commitment to sustainability and resilience will predictably lead to improved social and environmental outcomes.

Positive leadership and responsible stewardship of all resources lead to more acceptance of governmental services delivery and outcomes on resilience planning, which intensifies answerability, accountability, and good governance. There is a balance between climate resilience planning and overall municipal strategies and plans, achieving positive economic growth, social cohesion, environmental harmony, and good governance. As all the layers of the organization embrace resilient and sustainable strategies and policies, embedding them within structures, good governance, and culture change and transformation lead to improvements and micro-and macro-level solutions to climate-related issues and problems. The local government organizations have a crucial role in the policy development, program implementation, and the practical applications of climate resilience plans and strategies, including pandemic action plans and responses.

Outcomes, Discussions, and Further Considerations

- Define climate resilience, climate preparedness, climate mitigation, climate adaptation, resilience, and resilient and sustainable community.
- Analyze communities' plans to reduce greenhouse gas emissions and to address the climate change threats.
- Discuss the impact of climate change on the nation, state, and region.
- Assess the essential features for strategic climate preparedness and resilience planning in communities.
- Evaluate regional, local, state, and national levels of climate preparedness and resilience planning.
- Comprehend the relevance of the pandemic action plan in the emergency preparedness context.
- Examine synergies between private and public sectors initiatives to counter climate change and pandemics.

References

Agrawala S et al (2011) Private sector engagement in adaptation to climate change: approaches to managing climate risks. OECD environment working papers, no. 39. OECD Publishing. https://doi.org/10.1787/5kg221jkf1g7-en

Alibašić H (2017a) Measuring the sustainability impact in local governments using the Quadruple Bottom Line. Int J Sustain Policy Pract 13(3):37–45

Alibašić H (2017b) Sustainable energy policy for local and state governments. In: Farazmand A (ed) Global encyclopedia of public administration, public policy, and governance. Springer, Zurich

Alibašić H (2018a) Leading climate change at the local government level. In: Farazmand A (ed) Global encyclopedia of public administration, public policy, and governance. Springer, Cham. https://doi.org/10.1007/978-3-319-31816-5_3428-1

Alibašić H (2018b) Paris Climate Accord sans the USA: the role of local governments in confronting the effects of climate change and increasing resilience. Glob Stud J 11(2):33–39. https://doi.org/10.18848/1835-4432/CGP/v11i02/33-39

Alibašić H (2018c) The nexus of sustainability and climate resilience planning: embedding climate resilience policies in local governments. Int J Clim Change Impacts Responses 10(2):21–33. https://doi.org/10.18848/1835-7156/CGP/v10i02/21-33

Alibašić H (2018d) Role of corporations in addressing climate change. In: Farazmand A (ed) Global encyclopedia of public administration, public policy, and governance. Springer, Cham. https://doi.org/10.1007/978-3-319-31816-5_3429-1

Alibašić H (2020a) The administrative and ethical considerations of climate resilience: the politics and consequences of climate change. Public Integr. https://doi.org/10.1080/10999922.2020.1838142

Alibašić H (2020b) Sustainable procurement. In: Farazmand A (ed) Global encyclopedia of public administration, public policy, and governance. Springer, Cham. https://doi.org/10.1007/978-3-319-31816-5_3427-1

Alibašić H, Morgan JD (2020) Coastal climate readiness and preparedness: comparative review of the State of Florida and Cuba. In: Perception, design and ecology of the built environment: a focus on the global south. Springer. https://www.springer.com/gp/book/9783030258788

Baltimore Office of Sustainability (2013) Baltimore climate action plan. http://www.baltimoresustainability.org/wp-content/uploads/2015/12/BaltimoreClimateActionPlan.pdf

Baltimore Office of Sustainability (n.d.) Baltimore office of sustainability. https://www.baltimoresustainability.org/

Blaikie P, Cannon T, Davis I, Wisner B (1994/2014) At risk: natural hazards, people's vulnerability and disasters. Routledge, New York

Bloetscher F (2012) Protecting people, infrastructure, economies, and ecosystem assets: water management and adaptation in the face of climate change. J Water Water 4:367–388. https://doi.org/10.3390/w40x000x

Bloetscher F, Hoermann S, Berry L (2017) Adaptation of Florida's urban infrastructure to climate change. In: Chassignet EP, Jones JW, Misra V, Obeysekera J (eds)

Florida's climate: changes, variations, & impacts. Florida Climate Institute, Gainesville, pp 311–338. https://doi.org/10.17125/fci2017.ch11
Boswell MR, Greve AI, Seale TL (2012) Local climate action planning. Island Press, Washington, DC
Brady J (2018, July 11) The high cost of hot. Climate central research report. www.climatecentral.org/news/report-thehigh-cost-of-hot-21879
Broward County (2015) Climate action plan: local strategy to address climate change. http://www.broward.org/NaturalResources/ClimateChange/Documents/BrowardCAPReport2015.pdf
Broward County (2020) Broward County 2020 climate change action plan. https://www.broward.org/Climate/Documents/CCAP_2020_ADA.pdf
Broward County Government (2021a) Environmental protection and growth management. https://www.broward.org/resilience/Pages/default.aspx
Broward County Government (2021b) Resilience. https://www.broward.org/resilience/Pages/default.aspx
Broward County Government (2021c) Climate, energy & sustainability programs. https://www.broward.org/climate/Pages/default.aspx
Brugmann J (2012) Financing the resilient city. Environ Urban 24:215–232. https://doi.org/10.1177/0956247812437130
Brunetti C, Dennis B, Gates D, Hancock D, Ignelli D, Kiser EK, Kotta G, Kovner A, Rosen RJ, Tabor NK (2021, March 21). Climate change and financial stability. The Federal Reserve. https://www.federalreserve.gov/econres/notes/feds-notes/climate-change-and-financial-stability-20210319.htm
Bullock JA, Haddow GD, Haddow KS, Coppola DP (2014) Living with climate change: how communities are surviving and thriving in a changing climate. CRC Press, Chicago
Burch S (2010) In pursuit of resilient, low carbon communities: an examination of barriers to action in three Canadian cities. Energy Policy 38:7575–7585. https://doi.org/10.1016/j.enpol.2009.06.070
Byrne J, Hughes K, Toly N, Wang Y (2006) Can cities sustain life in the greenhouse? Bull Sci Technol Soc 26:84–95. https://doi.org/10.1177/0270467606287532
Chassignet EP, Jones JW, Misra V, Obeysekera J (eds) (2017) Florida's climate: changes, variations, & impacts. Florida Climate Institute, Gainesville. 10.17125/fci2017
City and County of San Francisco (2013) San Francisco climate action strategy. Department of Environment. https://sfenvironment.org/sites/default/files/engagement_files/sfe_cc_ClimateActionStrategyUpdate2013.pdf
City and County of San Francisco (2017) Climate plans and reports. San Francisco Environment. https://sfenvironment.org/climate-plans-reports
City and County of San Francisco (2021) Policy for climate change. https://sfenvironment.org/climate-change/policy
City of Ann Arbor (2012) City of Ann Arbor climate action plan. http://www.a2gov.org/departments/systems-planning/planning-areas/energy/Documents/CityofAnnArborClimateActionPlan_low%20res_12_17_12.pdf
City of Ann Arbor (2013) Sustainability framework. http://www.a2gov.org/departments/systems-planning/planning-areas/climate-sustainability/sustainability/Pages/SustainabilityFramework.aspx
City of Ann Arbor (2015) Sustainability action plan draft. http://www.a2gov.org/departments/systems-planning/planning-areas/climate-sustainability/Sustainability-Action-Plan/Documents/SAP%20-%20DRAFT%20-%20%20July%202015%20-%20web.pdf
City of Ann Arbor (2020a). Ann Arbor's Living Carbon Neutrality Plan. Retrieved from https://www.a2gov.org/departments/sustainability/Documents/A2Zero%20Climate%20Action%20Plan%20_3.0.pdf
City of Ann Arbor (2020b) Adaptation and resilience. https://www.a2gov.org/departments/sustainability/Adaptation-Resilience/Pages/default.aspx
City of Austin (2014) Report on climate resilience resolution 20131121-060; from Lucia Athens, Chief Sustainability Officer; Zach Baumer, Climate Program Manager. http://www.austintexas.gov/sites/default/files/files/Toward_a_Climate_Resilient_Austin.pdf
City of Austin (2015) Austin community climate plan. http://www.austintexas.gov/sites/default/files/files/Sustainability/FINAL_-_OOS_AustinClimatePlan_061015.pdf
City of Austin (2018) Climate resilience action plan for city assets and operations. https://austintexas.gov/sites/default/files/files/Sustainability/Climate_Resilience_Action_Plan.compressed.pdf
City of Baltimore (n.d.) Baltimore Office of Sustainability. https://www.baltimoresustainability.org/projects/climate-and-resilience/
City of Boston (2014) Greenovate Boston: 2014 climate action plan update. https://www.boston.gov/sites/default/files/greenovate_boston_2014_cap_update.pdf
City of Boston (2017) Resilient Boston: an equitable and connected city. https://www.boston.gov/sites/default/files/file/document_files/2017/07/resilient_boston.pdf
City of Chicago (2008) Climate Action Plan: out city, out future. http://www.chicagoclimateaction.org/filebin/pdf/finalreport/CCAPREPORTFINALv2.pdf
City of Chicago (2019) Resilient Chicago: a plan for inclusive growth and connected city. https://resilient.chicago.gov/download/Resilient%20Chicago.pdf
City of Chicago (n.d.) Sustain Chicago. https://sustainchicago.cityofchicago.org/
City of Detroit (2021) Detroit climate strategy. https://detroitmi.gov/departments/general-services-department/office-sustainability/detroit-climate-strategy
City of Dubuque (2013) Dubuque community climate action & resiliency plan 2013. Sustainable Dubuque. http://cityofdubuque.org/DocumentCenter/View/18359

City of Dubuque (2017) Sustainable Dubuque. http://www.sustainabledubuque.org/

City of Dubuque (2020) City of Dubuque climate action plan 2020. https://bloximages.newyork1.vip.townnews.com/telegraphherald.com/content/tncms/assets/v3/editorial/a/a3/aa3ed15f-0203-5148-aeda-d26e33972b55/5f0d18f0c9707.pdf.pdf

City of Eugene (2010) A community climate and energy action plan for Eugene. https://www.eugene-or.gov/Archive/ViewFile/Item/80

City of Eugene (2020) Eugene's community climate action plan 2.0: a roadmap for Eugene's climate journey. https://www.eugene-or.gov/DocumentCenter/View/55835/CAP-20_Summer_2020_FINAL-w-appendices-compressed

City of Eugene (n.d.) Climate recovery summary. https://www.eugene-or.gov/3210/Climate-Recovery-Summary

City of Fayetteville (n.d.) Arkansans can take steps to respond to climate change. http://www.fayetteville-ar.gov/DocumentCenter/View/14890

City of Los Angeles (2015a) The pLAn. Office of Mayor Eric Garcetti. http://plan.lamayor.org/wp-content/uploads/2017/03/the-plan.pdf

City of Los Angeles (2015b) Los Angeles climate action report: updated 1990 Baseline and 2013 emissions inventory summary. https://www.lamayor.org/sites/g/files/wph446/f/landing_pages/files/pLAn%20Climate%20Action-final-highres.pdf

City of Los Angeles (2018) Resilient Los Angeles. https://resilientcitiesnetwork.org/downloadable_resources/Network/Los-Angeles-Resilience-Strategy-English.pdf

City of New Orleans (2017) Climate action for a resilient New Orleans: 50% by 2030. Mayor's Office of Resilience and Sustainability. https://nola.gov/nola/media/Climate-Action/Climate-Action-for-a-Resilient-New-Orleans.pdf

City of New York (2017) 1.5°C: aligning New York City with the Paris climate agreement. Mayor's Office of Sustainability. https://www1.nyc.gov/assets/sustainability/downloads/pdf/1point5-aligning-nyc-with-paris-agreement.pdf

City of New York (2021) Mayor's Office of Climate Resiliency. https://www1.nyc.gov/site/orr/index.page

City of Orlando (2018) Orlando community action plan. https://www.orlando.gov/files/sharedassets/public/departments/sustainability/2018_orlando_communityactionplan.pdf

City of Pensacola (2018a) Climate Action recommendations: A blueprint for addressing climate change at the municipal level. https://www.cityofpensacola.com/DocumentCenter/View/20700/Climate-Mitigation-and-Adaptation-Final-Report?bidId=

City of Pensacola (2018b) Climate action recommendations a blueprint for addressing

City of Pensacola (n.d.) Climate mitigation and adaptation task force. https://www.cityofpensacola.com/3233/Climate-Mitigation-and-Adaptation-Task-F

City of Philadelphia (2016) Growing stronger: toward a climate-ready Philadelphia. https://beta.phila.gov/media/20160504162056/Growing-Stronger-Toward-a-Climate-Ready-Philadelphia.pdf

City of Philadelphia (2017) Publications. https://beta.phila.gov/departments/office-of-sustainability/publications/

City of Philadelphia (2021) Philadelphia climate action playbook. https://www.phila.gov/media/20210113125627/Philadelphia-Climate-Action-Playbook.pdf

City of Portland (2015) Climate action plan: local strategies to address climate change (Portland and Multnomah County 2015 Climate Action Plan). https://www.portlandoregon.gov/bps/article/531984

City of Portland and Mulnomanh County (2017) Climate action plan progress report. https://www.portlandoregon.gov/bps/article/636700

City of San Francisco (2021) San Francisco planning sustainability and resilience. https://sfplanning.org/resilience-and-sustainability

City of Tallahassee (2019) Tallahassee community resilience plan. https://talgov.com/uploads/public/documents/publicsafety/resplansm.pdf

City of Traverse City (2011) City of Traverse City Climate Action Plan. Solomon Townsend, Barton, Kirk and Michael Powers. SEEDS, Inc., Traverse City. http://www.traversecitymi.gov/downloads/climateactionplanfeb2011.pdf

Clearview Geographic LLC & Deady, E.L. (2021) Sea level rise vulnerability assessment: Pensacola, FL. https://storymaps.arcgis.com/stories/e812723f69ad4a618c8f5f8b08cb208e

Climate Central and ICF International (2015) States at risk: America's preparedness report card: Florida report. http://assets.statesatrisk.org/summaries/Florida_report.pdf

Climate Change at the Municipal Level (n.d.) Climate mitigation and adaptation task force City of Pensacola. https://www.cityofpensacola.com/DocumentCenter/View/15491/Climate-Mitigation-and-Adaptation-Task-Force-Report-PDF

Climate Mayors (2017) cities adopt the Parsi Climate Agreement goals. Retrieved from https://climatemayors.org/

County of Los Angeles (2015) Final unincorporated Los Angeles County Community Climate Action Plan (CCAP) 2020. Department of Regional Planning. http://planning.lacounty.gov/assets/upl/project/ccap_final-august2015.pdf

County of Los Angeles (2021) The Los Angeles County Community Disaster Resilience (LACCDR). http://www.laresilience.org/

Davenport C (2017) Counseled by industry, not staff, EPA chief is off to a blazing start. The New York Times. https://www.nytimes.com/2017/07/01/us/politics/trump-epachief-pruitt-regulations-climate-change.html

Desouzaa KC, Flaneryb TH (2013) Designing, planning, and managing resilient cities: a conceptual framework. Cities 35:89–99

Dong W, Liu Z, Liao H, Tang Q, Li X (2014) New climate and socioeconomic scenarios for assessing global human health challenges due to heat risk. Clim Change 130:505–518

Dzigbede KD, Gehl SB, Willoughby K (2020) Disaster resiliency of U.S. local governments: insights to strengthen local response and recovery from the COVID-19 pandemic. Public Administration Review 80(4). https://doi.org/10.1111/puar.13249

Environmental & Law Policy Center (2019) An assessment of the impacts of climate change on the Great Lakes. https://elpc.org/wp-content/uploads/2020/04/2019-ELPCPublication-Great-Lakes-Climate-Change-Report.pdf

Escambia County (2017) Florida community resiliency initiative pilot project adaptation plan for Escambia County, Florida May 2017. https://floridadep.gov/sites/default/files/CRI_Adaptation_Plan_Escambia_Appendix_May_2017.pdf

Fiksel J (2003) Designing resilient, sustainable systems. Environ Sci Technol 37:5330–5339. https://doi.org/10.1021/es0344819

Fiksel J (2006) Sustainability and resilience: toward a systems approach. Sustain Sci PractPolicy 2(2):14–21. http://ejournal.nbii.org

Fitzgerald J (2010) Emerald cities: urban sustainability and economic development. Oxford University Press, New York

Fitzgerald BG, O'Doherty T, Moles R, O'Regan B (2012) A quantitative method for the evaluation of policies to enhance urban sustainability. Ecol Indic 18:371–378. https://doi.org/10.1016/j.ecolind.2011.12.002

Fleming LN (2021, July 7) Ann Arbor mayor proposes 20-year climate change tax. The Detroit News. https://www.detroitnews.com/story/news/local/michigan/2021/07/07/ann-arbors-mayor-seeks-1-mill-tax-increase-fight-climate-change/7894927002/

Fletcher C (2013) Climate change: what the science tells us. Wiley, Heboken

Gimenez R, Hernantes J, Labaka L, Hiltz SR, Turoff M (2017) Improving the resilience of disaster management organizations through virtual communities of practice: a Delphi study. J Contingencies Crisis Manag 25:160–170

Gordon DJ (2016) The politics of accountability in networked urban climate governance. Glob Environ Politics 16(2):82–100

Griffin P (2017) The carbon majors database. CDP carbon majors report 201. 100 fossil fuel producers and nearly 1 trillion tonnes of greenhouse gas emissions. CDP Worldwide. https://b8f65cb373b1b7b15feb-c70d8ead6ced550b4d987d7c03fcdd1d.ssl.cf3.rackcdn.com/cms/reports/documents/000/002/327/original/Carbon-Majors-Report-2017.pdf?1499691240

Hallegatte S, Henriet F, Corfee-Morlot J (2011) The economics of climate change impacts and policy benefits at city scale: a conceptual framework. Clim Change 104:51–87. https://doi.org/10.1007/s10584-010-9976-5

Hillman J, Geunther E (2021) Organizational resilience: a valuable construct for management research? Int J Manag Rev 23:7–44. https://doi.org/10.1111/ijmr.12239

Intergovernmental Panel on Climate Change (2015) Climate change 2014: synthesis report. In: Contribution of Working Groups I, II and III to the Fifth Assessment Report of the Intergovernmental Panel on Climate Change. IPCC, Geneva. https://www.ipcc.ch/site/assets/uploads/2018/02/SYR_AR5_FINAL_full.pdf

Intergovernmental Panel on Climate Change (IPCC) (2013) Climate change 2013: the physical science basis. Contribution of Working Group I to the Fifth Assessment Report of the Intergovernmental Panel on Climate Change. Cambridge University Press, Cambridge, 1535p. https://doi.org/10.1017/CBO9781107415324

Intergovernmental Panel on Climate Change (IPCC) (2014) Summary for policymakers. In: Field Christopher B, Barros VR, Dokken DJ, Mach KJ, Mastrandrea MD, Bilir TE, Chatterjee M, Ebi KL, Estrada YO, Genova RC, Girma B, Kissel ES, Levy AN, MacCracken S, Mastrandrea PR, White LL (eds) Climate change 2014: impacts, adaptation, and vulnerability. Part A: global and sectoral aspects. Contribution of Working Group II to the Fifth Assessment Report of the Intergovernmental Panel on Climate Change. Cambridge University Press, Cambridge, pp 1–32

Intergovernmental Panel on Climate Change (IPCC) (2018) Global warming of 1.5 C: an IPCC special report on the impacts of global warming of 1.5 C above pre-industrial levels and related global greenhouse gas emission pathways, in the context of strengthening the global response to the threat of climate change, sustainable development, and efforts to eradicate poverty. Summary for policymakers. http://report.ipcc.ch/sr15/pdf/sr15_spm_final.pdf

Intergovernmental Panel on Climate Change (IPCC) (2021) Climate change 2021: the physical science basis. In: Masson-Delmotte V, Zhai P, Pirani A, Connors SL, Péan C, Berger S, Caud N, Chen Y, Goldfarb L, Gomis MI, Huang M, Leitzell K, Lonnoy E, Matthews JBR, Maycock TK, Waterfield T, Yelekçi O, Yu R, Zhou B (eds) Contribution of Working Group I to the Sixth Assessment Report of the Intergovernmental Panel on Climate Change. Cambridge University Press. In Press

Karl TR, Melillo JM, Peterson TC (eds) (2009) Global climate change impacts in the United States. US Global Change Research Program. https://downloads.globalchange.gov/usimpacts/pdfs/climate-impacts-report.pdf

Kling GW, Hayhoe LB, Johnson JJ, Magnuson S, Polasky SK, Robinson BJ, Shuter MM, Wander DJ, Wuebbles DR, Zak RL, Lindroth SC, Moser, Wilson ML (2003) Confronting climate change in the Great Lakes Region: impacts on our communities and ecosystems. Union of Concerned Scientists, Ecological Society of America, Cambridge, Washington, DC. https://mybb.gvsu.edu/bbcswebdav/pid-3951258-dt-content-rid-36059702_1/courses/GVPA671.01.201710/GVPA671.01.201330_

ImportedContent_20130423031707/greatlakes_final.pdf

Lindfield M (2010) Cities: a global threat and a missed opportunity for climate change. Environ Urban Asia 1(2):105–129. https://doi.org/10.1177/097542531000100202

Linnenluecke MK, Griffiths A (2012) Assessing organizational resilience to climate and weather extremes: Complexities and methodological pathways. Clim Change 113:933–947

Luber G et al (2014) Ch. 9: Human health. In: Melillo JM, Richmond T, Yohe GW (eds) Climate change impacts in the United States: the third national climate assessment. U.S. Global Change Research Program, pp 220–256. https://doi.org/10.7930/J0PN93H5

Mach KJ, Mastrandrea MD, Bilir E, Field CB (2016) Understanding and responding to danger from climate change: the role of key risks in the IPCC AR5. Clim Change 136:427–444. https://doi.org/10.1007/s10584-016-1645-x

Malcolm JR, Liu C, Neilson RP, Hansen L, Hannah L (2006) Global warming and extinctions of endemic species from biodiversity hotspots. Conserv Biol 20:538–548

Mann ME, Peterson TC, Hassol SJ (2017a) What we know about the climate change-hurricane connection. Scientific American. https://blogs.scientificamerican.com/

Mann ME, Lloyd EA, Oreskes N (2017b) Assessing climate change impacts on extreme weather events: the case for an alternative (Bayesian) approach. Clim Change 144(2):131–142. https://doi.org/10.1007/s10584-017-2048-3

Martin County (2021) Sea level rise report 2021 impact & analysis. https://view.publitas.com/martin-county-board-of-county-commissioners/martin-county-sea-level-rise-report-2021/page/1

McGregor J (2017) 'Climate change is real': CEOs share their disappointment over Trump's Paris accord exit. The Washington Post. https://www.washingtonpost.com/news/on-leadership/wp/2017/06/01/ceos-make-final-pleas-to-trump-to-stay-in-paris-climate-agreement/?utm_term=.5fbbcaa4cbda

Megginson A (2018, Dec 7) City of Pensacola takes action against climate change. WUWF. https://www.wuwf.org/science/2018-12-07/city-of-pensacola-takes-action-against-climate-change

Miami-Dade County (2021) Strengthening resilience in Miami-Dade County. https://www.miamidade.gov/global/economy/resilience/home.page

Michigan Climate Action Network (MCAN) (2016) Traverse City commits to 100% clean energy by 2020. http://www.miclimateaction.org/traverse_city_commits_to_100_percent_clean_energy_by_2020

Morrison J (2019, June 10) What will happen to Pensacola's climate change report? InWeekly. https://ricksblog.biz/what-will-happen-to-pensacolas-climate-change-report/

Morrison J (2020, Dec 10) Pensacola warms up to climate report (but, slowly) – update. InWeekly. https://ricksblog.biz/pensacola-warms-up-to-climate-report-but-slowly/

O'Brien KL, Leichenko RM (2000) Double exposure: assessing the impacts of climate change within the context of economic globalization. Glob Environ Chang 10(2000):221–232

Orange County (2021) Sustainable operations and resilience action plan. https://www.ocfl.net/portals/0/Library/Board%20Of%20County%20Commissioners/docs/1-19-21_OCFL%20Sustainable%20Operations%20Resilience_Action_Plan.pdf

Pecl G, Araújo M, Bell J, Blanchard J, Bonebrake T, Chen I-C et al (2017) Biodiversity redistribution under climate change: Impacts on ecosystems and human well-being. Science 355(6332). https://doi.org/10.1126/science.aai9214

Powers SD (2015) City of Ann Arbor Administrator General Fund Budget Memo. http://media.mlive.com/annarbornews_impact/other/city_admin_FY16_budget.pdf

Resilient Communities for America (2015) Resilient communities for America. http://www.resilientamerica.org/

Resilient Macomb (2016) Planning for coastal resilience in Macomb County, Michigan. http://www.resilientmichigan.org/downloads/macomb_coastalresiliencyreport_web.pdf

Resilient Michigan (n.d.) Resilient Michigan communities. http://www.resilientmichigan.org/communities.asp

Resilient Monroe (2013) Resilient Monroe resource atlas. http://www.resilientmichigan.org/downloads/resilient_monroe_resource_atlas_20131015.pdf

Riley T (2017) Just 100 companies responsible for 71% of global emissions, study says. The Guardian. Guardian Sustainable Business. https://www.theguardian.com/sustainable-business/2017/jul/10/100-fossil-fuel-companies-investors-responsible-71-global-emissions-cdp-study-climate-change

Segan DB, Hole DG, Donatti CI, Zganjar C, Martin S, Buthart SHM, Watson JEM (2015) Considering the impact of climate change on human communities significantly alters the outcome of species and site-based vulnerability assessments. Divers Distrib 21:1101–1111

Silva RA, West JJ, Zhang Y, Anenberg SC, Lamarque JF, Shindell DT, Collins WJ et al (2013) Global premature mortality due to anthropogenic outdoor air pollution and the contribution of past climate change. Environ Res Lett 8(034005):1–11. https://doi.org/10.1088/1748-9326/8/3/034005

Solomon H, Kopp RE (2018) An economist's guide to climate change science. J Econ Perspect 32(Fall):3–32

Solomon H, Kopp RE, Jina A, Rising J, Delgado M, Shashank Mohan DJ, Rasmussen RM-W, Wilson P, Oppenheimer M, Larsen K, Hauser T (2017) Estimating economic damage from climate change in the United States. Science 356(June):1362–1369

Son C, Sasangohar F, Peres CS (2017) Redefining and measuring resilience in emergency management systems. Hum Fact Ergon Soc 61(1). https://doi.org/10.1177/1541931213601899
Southeast Florida Regional Climate Change Compact Counties (2012) Regional climate action plan. A region responds to a changing climate. https://southeastfloridaclimatecompact.files.wordpress.com/2014/05/regional-climate-action-plan-final-ada-compliant.pdf
Southeast Florida Regional Climate Change Compact Counties (2021) Advancing resilience solutions through regional action. https://southeastfloridaclimatecompact.org/
Stanton R (2015) See details of Ann Arbor's proposed $380M city budget. Mlive. http://www.mlive.com/news/ann-arbor/index.ssf/2015/04/ann_arbors_proposed_380m_city.html
Stott PA, Christidis N, Otto FEL, Sun Y, Vanderlinden J, van Oldenborgh GK et al (2016) Attribution of extreme weather and climate-related events. Wiley Interdiscip Rev Clim Change 7(1):23–41. https://doi.org/10.1002/wcc.380
Tesla N (1900) The problem of increasing human energy: with special reference to the harnessing of the sun's energy. The Century Magazine. http://www.teslauniverse.com/nikola-tesla-article-the-problem-of-increasing-human-energy
The State, Local, and Tribal Leaders Task Force on Climate Preparedness and Resilience (Task Force) (2014) Recommendations to the President. https://obamawhitehouse.archives.gov/sites/default/files/docs/task_force_report_0.pdf
The United States Conference of Mayors (USCM) (2005) Mayors climate protection agreement. https://www.usmayors.org/mayors-climate-protection-center/
The United States Conference of Mayors (USCM) (2017) 85th Annual meeting 2017 adopted resolutions. http://legacy.usmayors.org/resolutions/85th_Conference/proposedcommittee.asp?committee=Energy
The United States Department of Energy (DOE) (2013) U.S. energy sector vulnerabilities to climate change and extreme weather. https://energy.gov/sites/prod/files/2013/07/f2/20130716-Energy%20Sector%20Vulnerabilities%20Report.pdf
The United States Environmental Protection Agency (EPA) (2018) Green power partnership. Green Power Partnership Top Partner Rankings. https://www.epa.gov/greenpower/green-power-partnership-top-partner-rankings
The White House (2014) President Barack Obama. President's state, local, and tribal leaders task force on climate preparedness and resilience. Recommendations to the President. https://obamawhitehouse.archives.gov/sites/default/files/docs/task_force_report_0.pdf
The White House (2021, April 22) FACT SHEET: president biden sets 2030 greenhouse gas pollution reduction target aimed at creating good-paying union jobs and securing U.S. Leadership on Clean Energy Technologies. https://www.whitehouse.gov/briefing-room/statements-releases/2021/04/22/fact-sheet-president-biden-sets-2030-greenhouse-gas-pollution-reduction-target-aimed-at-creating-good-paying-union-jobs-and-securing-u-s-leadership-on-clean-energy-technologies/
The World Bank (2012) Turn down the heat: why a 4°C warmer world must be avoided. A report for the World Bank by the Potsdam Institute for Climate Impact Research and Climate Analytics. https://mybb.gvsu.edu/bbcswebdav/pid-3951253-dt-content-rid-36059696_1/courses/GVPA671.01.201710/GVPA671.01.201330_ImportedContent_20130423031707/Turn_Down_the_heat_Why_a_4_degree_centrigrade_warmer_world_must_be_avoided.pdf
Tveiten CK, Albrechtsen E, Wærø I, Wahl AM (2012) Building resilience into emergency management. Safety Sci 50:1960–1966. https://doi.org/10.1016/j.ssci.2012.03.001
U.S. Department of State (2017) Communication regarding intent to withdraw from Paris Agreement. Office of the Spokesperson. https://www.state.gov/r/pa/prs/ps/2017/08/273050.htm
U.S. Global Change Research Program (USGCRP) (2017) In: Wuebbles DJ, Fahey DW, Hibbard KA, Dokken DJ, Stewart BC, Maycock TK (eds) Climate science special report: fourth national climate assessment, vol I. U.S. Global Change Research Program, Washington, DC, 470p. https://doi.org/10.7930/J0J964J6
U.S. Securities and Exchange Commission (2021, March 15) Speech: a climate for change: meeting investor demand for climate and ESG information at the SEC. https://www.sec.gov/news/speech/lee-climate-change
United Nations Framework Convention on Climate Change (UNFCCC) (2015) Adoption of the Paris Agreement. United Nations Framework Convention on Climate Change. https://unfccc.int/resource/docs/2015/cop21/eng/l09r01.pdf
United State Census Bureau (2021) QuickFacts Pensacola city, Florida. https://www.census.gov/quickfacts/fact/table/pensacolacityflorida/PST045219
United States Census Bureau (2019) Quick facts United States. https://www.census.gov/quickfacts/fact/table/US/VET605219
United States Census Bureau (2021) QuickFact Broward County, Florida. https://www.census.gov/quickfacts/fact/table/browardcountyflorida/PST045219
United States Environmental Protection Agency (US EPA) (2021a) Green power partnership top 30 local government. https://www.epa.gov/greenpower/green-power-partnership-top-30-local-government
United States Environmental Protection Agency (US EPA) (2021b) Green Power Partnership Fortune 500® Partners List. https://www.epa.gov/greenpower/green-power-partnership-fortune-500r-partners-list
United States Environmental Protection Agency (US EPA) (2021c) Green Power Partnership 100% Green Power Users. https://www.epa.gov/greenpower/green-power-partnership-100-green-power-users

Urban MC (2015) Accelerating extinction risk from climate change. Science 348:571–573
Välikangas L, Li PP (2020) Resilience, resilience, resilience. Manage Org Rev 16(2):225–227. https://doi.org/10.1017/mor.2020.18
Van den Oord S, Vanlaer N, Marynissen H, Brugghemans B et al (2020) Network of networks: preliminary lessons from the Antwerp Port Authority on crisis management and network governance to deal with the COVID-19 pandemic. Public Administr Rev 80(4):880–894. https://doi.org/10.1111/puar.13256
Visser H, Petersen AC, Ligtvoet W (2014) On the relation between weather-related disaster impacts, vulnerability and climate change. Clim Change 125:461–477
Walsh J et al (2014) Ch. 2: Our changing climate. In: Melillo JM, Richmond T, Yohe GW (eds) Climate change impacts in the United States: the third national climate assessment. U.S. Global Change Research Program, pp 19–67. https://doi.org/10.7930/J0KW5CXT
Yongyang C, Lontzek TS (2019) The social cost of carbon with economic and climate risks. J Polit Econ 127(December):2684–2734
Zautra A, Hall J, Murray K (2008) Community development and community resilience: an integrative approach. Commun Dev. 39(3). https://doi.org/10.1080/15575330809489673

7 Evaluating Tools and Resources for Strategic Resilience and Sustainability Planning

"Furthermore, towns and cities with their monuments, vast constructions, and large buildings, are set up for the masses and not for the few. Therefore, united effort and much co-operation are needed for them."

Ibn Khaldun (1406, 1969). The Muqaddimah: An introduction to history; In three volumes.

Overview

The seventh chapter delivers an illustrative review of available resources, tools, and assistance for strategic resilience and sustainability planning to accomplish maximum organizational resilience and sustain communities. Chapter seven reviews the resources available for the strategic resilience and sustainability plans. Beyond consulting services, resources and tools are available at no or at an economically feasible cost that requires internal coordination, staffing, and management. Many organizations opt for more feasible in-house completion of plans, including greenhouse gas emission reports. The chapter provides a brief overview of accessible national, state level, and regional resources and organizations supporting the climate preparedness and resilience efforts, including available tools for carbon footprint inventory and greenhouse gas emissions reporting. An illustrative template of strategic resilience and sustainability plan is provided. A template for integrating climate readiness into emergency preparedness and disaster mitigation plan is recommended, and responses to a pandemic are analyzed.

Keywords

Tools · Climate tools · Carbon footprint Sustainability planning · Resilience planning Carbon inventory · Climate preparedness Climate resilience · Emergency preparedness Carbon footprint · Climate readiness · Climate change · Pandemic

Key Questions

The goal of the seventh chapter of the book is to answer the following underlying assumptions and questions.

- What resilience, sustainability, and climate preparedness resources do local government organizations need to plan for resilient and sustainable communities and to achieve maximum organizational resilience?
- What tools and data do communities necessitate for climate preparedness and resilience?
- What tools and resources are available for resilience planning?
- What carbon inventory and forecasting resources are recommended for the organizations?

H. Alibašić, *Strategic Resilience and Sustainability Planning*, Sustainable Development Goals Series, https://doi.org/10.1007/978-3-030-91159-1_7

- What templates or outlines are appropriate for cities' resilience and sustainability planning?
- How can communities integrate climate readiness into emergency preparedness?
- How can organizations integrate pandemic action plans into disaster readiness and responses?
- How can private sector firms contribute to climate resilience and preparedness in communities?

Introduction

The seventh chapter of the book provides a perfunctory and illustrative review of available resources, tools, support for resilience and sustainability planning for counties, cities, and other organizations to attain maximum resilience and sustain communities. Beyond private and often expensive consulting services, resources, and tools are available at no cost to organizations, and require internal coordination, staffing, and management.

The review is intended as a possible first step for organizations interested in starting resilience and sustainability strategic planning. It is not an inclusive review as there are abundant resources, agencies, and tools available to local governments, nonprofit, and private sector organizations regionally, in states, or nationally and internationally. An illustrative overview and a template of organizational resilience and sustainability plan using the Quadruple Bottom Line elements are provided. A template for integrating climate readiness for heatwaves and wildfires into emergency management and disaster preparedness and mitigation plans is presented.

Resilience and Sustainability Planning Organizations and Resources

A brief overview of several national, regional, and statewide resources and organizations available to assist in strategic resilience and sustainability planning is provided. Federal and state agencies, national and international organizations offer resources and tools for resilience, climate preparedness, disaster planning, and sustainability planning.

For instance, the US Environmental Protection Agency Pollution Prevention (US EPA P2) (2021a) program website offers information for practices to reduce, eliminate, or prevent pollution before recycling, treatment, or disposal.

Federal Agencies Resilience Resources

The United States Environmental Protection Agency P2 Greenhouse Gas Emissions Calculator

The commitment to reducing the harmful impacts of greenhouse gas emissions is essential in resilience and sustainability planning. The resources and tools provided by the United States Environmental Protection Agency are particularly effective for organizations that may lack resources to retain consultants and rely on internal staff to measure negative impacts of operations. The website includes case studies and measuring pollution prevention (US EPA 2020). The P2 GHG emissions calculator quantifies and inventories annual greenhouse gas (GHG) emissions and translates projects into pollution reduction to enhance resilience. The excel sheet tool is divided into tabs and calculates and aggregates CO_2 emissions, costs savings from emissions reduction, power management, green energy use, green chemistry, mobility, transportation, and water management (US EPA 2020). The P2 cost calculator's unique features convert the greenhouse gas emissions reductions to related cost savings, drawing on data in the P2 Cost Savings Calculator (US EPA 2020).

The climate change resources were removed from the EPA website in 2017 during the previous administration, and the cities of Houston and Milwaukee preserved and housed decades of research on their websites (City of Houston 2017; City of Milwaukee 2017). The US EPA (2021b) acknowledged that the previous administration forcefully removed climate change information, stating that "EPA's climate change website is back, with more content to come," adding that "understanding and addressing climate change is critical to EPA's mission of protecting human health and the environment" (para. 1 and 2). The US EPA (2021b) "tracks and reports greenhouse gas emissions," leveraging rigorous science, "to reduce emissions to combat climate change" (para. 2).

The White House

Climate readiness for various levels of government was furthered by President Obama's State, Local and Tribal Leaders Task Force on Climate Preparedness and Resilience (The White House 2014). In a sharp reversal of policies from his predecessor, President Biden made it a top priority to combat the threats of climate crises. The White House (2021b) under President Biden rejoined the Paris Climate Agreement, ordered all federal agencies to ensure science is reintroduced back into climate change threat analysis and examinations. President Biden directed federal agencies to reengage on climate change to correct the lack of engagement in the past to advance environmental equity and justice, public health improvements, environmental protection, reduce greenhouse gas emissions, to rely on science and bolster resilience to ensure integrity in decision making (The White House 2021a, b, c). Furthemore, President Biden directed "all executive departments and agencies to immediately review and, as appropriate and consistent with applicable law, take action to address the promulgation of Federal regulations and other actions during the last four years that conflict with these important national objectives, and to immediately commence work to confront the climate crisis" (The White House 2021a, c, para 2 and 3). It requires that "the heads of all agencies shall immediately review all existing regulations, orders, guidance documents, policies, and any other similar agency actions (agency actions) promulgated, issued, or adopted between January 20, 2017, and January 20, 2021, that are or maybe inconsistent with, or present obstacles to the policy outlined in section 1" of the order (The White House 2021a, c, para 2 and 3). Among several steps, resilient strategies will include: Reducing Methane Emissions in the Oil and Gas Sector; Establishing Ambitious, Job-Creating Fuel Economy Standards; Job-Creating Appliance- and Building-Efficiency Standards; Protecting Our Air from Harmful Pollution; and Accounting for the Benefits of Reducing Climate Pollution (The White House 2021a, c, para 2 and 3).

The important identifier of the true cost of climate change is the introduction of the methodology to calculate the various costs of greenhouse gas emissions. President Biden's ordered agencies to capture "the full costs of greenhouse gas emissions as accurately as possible, including by taking global damages into account," to facilitate "sound decision-making, recognize the breadth of climate impacts and support the international leadership of the United States on climate issues" (The White House 2021a). As noted in the order "the social cost of carbon (SCC), the social cost of nitrous oxide (SCN) and social cost of methane (SCM) are estimates of the monetized damages associated with incremental increases in greenhouse gas emissions," and "are intended to include changes in net agricultural productivity, human health, property damage from increased flood risk, and the value of ecosystem services" (The White House 2021a, para 22). The order is accurate in noting that "an accurate social cost is essential for agencies to accurately determine the social benefits of reducing greenhouse gas emissions when conducting cost-benefit analyses of regulatory and other actions" (The White House 2021a, para. 22). An accurate account of greenhouse gas emissions is crucial to combining the precise measurements of variables introduced into the governance, economic, environmental, and social factors occurring in society due to climate change, and the timing and precision are necessary to account for all these threats and costs.

National Aeronautics and Space Administration (NASA)

The National Aeronautics and Space Administration (NASA) is committed to addressing climate change and promoting climate resilience and sustainability (NASA 2021a, b, c). NASA's (2021a, para. 4) sustainability objectives include: increasing energy efficiency and the use of renewable energy; measuring, reporting, and reducing NASA's direct and indirect greenhouse gas emissions; conserving and protecting water resources through efficiency, reuse, and stormwater management; eliminating waste, preventing pollution, and increasing recycling; leveraging agency purchasing power to foster markets for sustainable technologies and environmentally preferable materials, products, and services; designing, constructing, maintaining, and operating high-performance, sustainable buildings; utilizing power management options, and reducing the number of the agency data centers; supporting economic growth and livability of the communities where NASA conducts business; evaluating agency climate change risks and vulnerabilities and develop mitigation and adaptation measures to manage both the short-and long-term effects of climate change on the agency's mission and operations; raising employee awareness and encourage each individual in the NASA community to apply the concepts of sustainability to every aspect of their daily work to achieve these goals; maintaining compliance with all applicable federal, state, local, or territorial law and regulations related to energy security, a healthy environment, and environmentally sound operations; and complying with internal NASA requirements and agreements with other entities (para. 4). According to the National Aeronautics and Space Administration (NASA) (2021b) organization consider itself to be "a world leader in climate studies and Earth science," and "while its role is not to set climate policy or prescribe particular responses or solutions to climate change, its purview does include providing the robust scientific data needed to understand climate change." Moreover, NASA (2021b) "makes this information available to the global community—the public, policy- and decision-makers and scientific and planning agencies around the world" (para. 1). NASA (2021b) defined mitigation as dropping and steadying the levels of heat-trapping greenhouse gases in the atmosphere and adaptation as adapting to the climate change already underway.

National Oceanic and Atmospheric Administration (NOAA)

The National Oceanic and Atmospheric Administration (NOAA) contains a US climate resilience planning tool with case studies comprising of the climate explorer, funding opportunities for resilience, and topics on the built environment, coasts, food, marine, energy, health, transportation, water, and tribal nations (NOAA n.d.). The NOAA's Climate.gov (n.d) website includes data snapshots and maps with 30-year precipitation by month, drought, temperatures, projections, severe weather, oceans, outlooks, climate dashboard, and climate variability.

Also, the NOAA's website features other organizations' resources and tools on climate change. Each tool serves as a guide to assist the organizations and communities with the various aspects of climate preparedness. The tools include Coastal Flood Exposure Mapper, Quick Report Tool for Socioeconomic Data, access to economic and demographic data for multiple coastal jurisdictions, and other valuable resources (NOAA 2017). The NOAA's Great Lakes Environmental Research Laboratory observes, monitors, and forecasts accurate and up-to-date Great Lakes water levels (NOAA 2018; Great Lakes Climate Resilience 2013).

United States Global Change Research Program (USGRCB)

The United States Global Change Research program (USGCRP) incorporates federal research on climate change effects and implications for society. The USGCRP (n.d.) program was created by Presidential initiative in 1989 and mandated by Congress in the Global Change Research Act (GCRA) of 1990, to advance a research program to "assess, predict, and respond to human-induced and natural processes of global change" (para. 1). It includes thirteen federal agencies that conduct or use research on climate change and its impacts on society. It functions under the direction of the Subcommittee on Global Change Research of the National Science and Technology Council's Committee on Environment.

The USGRCP (2017, 2018a, 2018b) produces and publishes National Climate Assessments, and the reports are available online (Mellilo et al. 2014). Other reports produced and published by the USGRCP (2016, 2018b) are the Impacts of Climate Change on Human Health in the United States and The Second State of Carbon Cycle Report (SOCCR2). The SOCCR2 is a decade-long appraisal of the carbon cycle in North America, with input from over 200 experts from national laboratories, universities, private sector, and research institutions and governments in Mexico, United States, and Canada (USGRCP 2018b). For the United States Global Change Research Program (USGCRP), Brown et al. (2015) reported on global food systems.

United States Fish & Wildlife Service

The United States Fish & Wildlife Service (2021a) acknowledged the enormous negative impact of climate change on wildlife, noting that "some populations may decline, many will shift their ranges substantially, and still, others will face [an] increased risk of extinction." Furthermore, the U.S. Fish & Wildlife Service (2021a) reported that "some species will survive in the wild only through direct and continuous intervention by wildlife and fisheries managers," and initiatives to counter climate change through series of strategic plan steps, including:

> "Developing expertise in biological carbon sequestration — sequestering greenhouse gases in plant biomass while also creating or restoring priority native plants, fish, and wildlife habitats — and foster efforts to sequester carbon on lands it manages," and "facilitating habitat conservation through carbon sequestration at the international level. By working with international partners and stakeholders to help reduce deforestation rates in key areas, such as tropical forests, the Service will help preserve areas critical to biodiversity conservation and support greenhouse gas mitigation" (para. 1 and 8). Moreover, "the rapid warming of the earth's atmosphere poses historical challenges for the world — and the National Wildlife Refuge System" (United States Fish & Wildlife Service 2021b, para. 1).

Transnational, National, State, and Regional Groups and Organizations Supporting Climate Resilience and Sustainability Initiatives

Intergovernmental Panel on Climate Change (IPCC)

With the issuance of the most recent climate change report, the IPCC (2021a), the working group of the sixth assessment report developed an interactive atlas. According to IPCC (2021b), the atlas is "A novel tool for flexible spatial and temporal analyses of much of the observed and projected climate change information underpinning the Working Group I contribution to the Sixth Assessment Report, including regional synthesis for Climatic Impact-Drivers (CIDs)." It invites participants to participate in user testing survey and to report any errata. Also, the NASA Sea Level Projection Tool permits "users to visualize and download the sea level projection data from the IPCC 6th Assessment Report (AR6)" (IPCC 2021c).

United Nations Framework Convention on Climate Change (UNFCCC) (2021)

The United Nations Framework Convention on Climate Change, with 197 members, is the United Nations (UN) agency assisting the worldwide responses to climate change threats (UNFCCC 2021a, b). It is charged to organize the annual UN Conference of the Parties (COP) climate change conferences, including the COP26, that was held in Glasgow, Scotland, in 2021 (UNFCCC 2021b). The UNFCCC (2021c) annual report acknowledges the difficulties of the pandemic in 2020 and its effects on combating climate change threats. The scope of the report included intergovernmental processes and the initiatives in 2020, supporting the Convention adopted in 1992, the Kyoto Protocol, and the 2015 Paris Climate Agreement. The report emphasizes the need for financing mitigation and adaptation initiatives.

Resilient Cities Network

The Rockefeller Foundation funded the 100 Resilient Cities program. Cities in the network were provided the financial and logistical support to create a position of a Chief Resilience Officer, with expertise to develop a "robust resilience strategy," and access to resources for the implementation of resilience strategies and information sharing between member cities (The Rockefeller Foundation 2021). With the 100 Resilient Cities grants, 1,000 cities around the globe applied for the support, and 100 communities were selected to become the Resilient Cities and hired Chief Resilience Officers (CROs) to start plans for climate resilience (The Rockefeller Foundation 2021). The 100 Chief Resilience Officers is an initiative aimed at building infrastructure to plan for climate change through climate resilience policies and programs. By assisting communities to employ a full staff person dedicated to climate resilience, more cohesive efforts are being deployed at subnational levels, allowing a community-based initiative to take hold. The 100 Resilient Cities initiative funding concluded in 2019, with 50 resilient strategies already adopted around the globe (The Rockefeller Foundation 2021). As a renewed attempt to support resilient cities, member cities, and Chief Resilience Officers from the 100 Resilient Cities program established the Resilient Cities Network "sharing a common lens for holistic urban resilience and with thousands of projects in implementation" (Resilient Cities Network 2021). The initiatives promoted by the network include water, waste management, recovery, energy, transportation, and urban oceans (Resilient Cities Network 2021).

The United States Conference of Mayors and National League of Cities

The Alliance for a Sustainable Future (2020), a collaborative effort by the United States Conference of Mayors and the Center for Climate and Energy Solutions (C2ES) issued a report on the state of climate change efforts by cities, showing significant climate resilience leadership by the US mayors. The United States Conference of Mayors through Mayors' Climate Protection Center and other resilience and sustainability initiatives encourage member cities and mayors to take a leadership role in addressing the threats of climate change and to meeting carbon reduction goals (USCM 2021). Similarly, the National League of Cities encourages its members to be proactive on climate resilience, environmental issues, sustainability, and climate change, and offers leadership training and assistance on climate issues (NLC 2017, 2021). Resilient communities are on the front lines of the climate change challenges.

Fiscal constraints and the impact of climate change and the recent pandemic on governance and the ability to deliver outcomes causes significant disruptions. Cities and counties engaged in climate preparedness and resilience efforts adapt and transform and accept discontinuities as they continue to offer services without interruptions. In a crisis, community-resilience related efforts are an opportunity and a tool for governments to change priorities, emergency management processes, and outcomes of the budget process. The Compact of Mayors (n.d.) and the Global Covenant of Mayors for Climate and Energy

(2021) support cities with set targets and goals in reducing carbon emissions.

Resilient Organizations and Initiatives Spotlight

The Great Lakes Cities Saint-Lawrence Cities Initiative offers networking opportunities, best practices, advocacy, and an array of tools and documents for its members on an array of issues related to resilience, sustainability, and energy (GLSLCI 2021a, b). Also, the initiatives for its members include resources and training on climate change adaptation and mitigation, with a focus on climate-ready cities toolkit, climate-ready infrastructure and strategic sites protocols, and other climate resilience tools (GLSLCI 2021b, c).

A non-profit organization, **Architecture 2030** (2018a), was established in 2002 to adjust the cities' built environment from being the main contributors of greenhouse gases to those addressing climate crises. It oversees 2030 Districts, the private–public challenge for energy consumption and carbon reduction in cities in the United States and Canada, including Seattle, Cleveland, Los Angeles, Pittsburgh, Denver, San Francisco, Dallas, Toronto, Albuquerque, San Antonio, Austin, and others. The commercial building owners and facility managers of government buildings in those cities are committed to reducing greenhouse gas emissions, water conservation, and energy consumption. 2030 Challenge issued to building owners include goals to establish a greenhouse gas emission baseline and targets, applying passive design strategies to achieve maximum energy efficiency and integrating energy-efficient technology and systems, and incorporating renewable energy to meet the buildings' energy demands (Architecture 2030 2018b).

ICLEI-Local Government for Sustainability (ICLEI) A global network of over 1,750 governments, ICLEI represents one of the most extensive networks of local governments. Since 1991 ICLEI has been providing climate resilience and sustainability support advice to local government leaders and administrators (ICLEI USA 2021a). In addition to offering ClearPath greenhouse gas emissions inventory and greenhouse gas emission protocol as part of the membership, ICLEI has additional tools for its members to prepare for climate change impacts and emissions management (ICLEI USA 2021b).

Urban Sustainability Directors Network (USDN) (2021) is a network of sustainability officers from cities around the United States sharing the best practices and information. It includes opportunities for collaborative projects.

The Nature Conservancy features links to the climate adaptation and planning tools, sustainability and resilience tools with geospatial data information (the Nature Conservancy 2021a, b).

Northern Gulf of Mexico Sentinel Site Cooperative (2021a, b) with its Gulf Tools for Resilience Exploration Engine (Gulf TREE) assists the government planners and administrators with climate preparedness and readiness planning. It is designed as a decision-support tree framework assisting users in the Gulf of Mexico region to identify the appropriate climate resilience tool used to evaluate and analyze the science-based data to prepare for hazards and resilience for the coastal areas. The Gulf TREE climate change and resilience exploration engine is a collaborative endeavor between the Northern Gulf of Mexico Sentinel Site Cooperative (SSC), the Gulf of Mexico Climate and Resilience Community of Practice (CoP), and the Gulf of Mexico Alliance (GOMA) team (Northern Gulf of Mexico Sentinel Site Cooperative 2022).

Resilient Michigan Resilient Michigan is a collaborative effort "developed by the Land Information Access Association (LIAA), non-profit community service and planning organization headquartered in Traverse City,

Michigan" (Resilient Michigan n.d.). According to its website, Planning for Resilient Communities advances "community-wide planning efforts that promote community resilience in the face of rapid economic changes and increasing climate variability" (Resilient Michigan n.d.). It published a practical handbook and resource guide with the following nine themes of resilience: Local Governance and Leadership; Gray and Green Infrastructure; Transportation; Local Food and Food Systems; Housing and Neighborhoods; Natural Resources; Public Health; Coastal Processes; and Energy (Resilient Michigan 2017).

Public Sector Perspective

Private companies utilize the P2 tools and are featured in the P2 case studies, including initiatives in energy efficiency and savings, water conservations, lifecycle assessment, waste reduction, and water reduction in production (Xu et al. 2016; University of Wisconsin-Madison 2016; Hardcastle 2016a). Ford and Pepsi Co. committed to saving water and, in the process, saving millions of dollars in operations (Hardcastle 2016a, b). The US EPA (2021c) provides a comprehensive recycling guide for businesses to help them reduce waste and reuse materials, including a recycling purchasing guide. As noted on the US EPA (2021c) website, "the Comprehensive Procurement Guideline (CPG) program is part of EPA's Sustainable Materials Management initiative that promotes a system approach to reducing materials use and the associated environmental impacts over the materials entire life cycle," and is "authorized by Congress under Section 6002 of the Resource Conservation and Recovery Act (RCRA) (42 United States Code 6962)" (para. 1 and 2). As climate change threats increase, large scale changes in political and policy issues such as climate change and pandemics require business and corporate leadership and partnership. Murray (2019) discussed the businesses involved in addressing climate change threats and those businesses on the opposite spectrum. It will take all sectors to contribute to the efforts to reduce greenhouse gas emissions to eventually alter the negative effects of climate change (Alibašić 2018, 2020; Davis et al. 2018). Large international corporations, including Google, Facebook, IKEA, Sida, Strantec, Iberdrola, Microsoft, BNP Paribas, partner with the United Nations on initiatives to counter climate change risks (UNFCCC 2021d).

Strategic Resilience and Sustainability Plan Outline

Each community and organization has a differing set of priorities and initiatives, and has different governance, economic, environmental, and social circumstances. However, the resilience and sustainability strategic plan consists of the following elements:

- Establishing categorical goals, outcomes, targets, and initiatives, in alignment with the vision, objectives, and mission of an organization or a community.
- Outlining the deadlines for accomplishing resilience and sustainability-related goals.
- Ascertaining the internal and external stakeholders, collaborators, partners, and supporters.
- Categorizing resilient and sustainable projects with the most desirable return on investment community-wide and organizationally, including governance, accountability, reporting, management and material use minimization, reuse, repurposing and recycling of materials, emergency management and disaster readi-

ness operations, building efficiency, energy use, renewable energy production, affordable housing projects, equity, justice, income disparity, and sustainable development.
- Integrating targets in the plan.
- Assigning outcome champions and implementers.
- Formulating greenhouse gas emissions and carbon footprint reduction and climate resilience objectives and targets.
- Defining measurements of the Quadruple Bottom Line outcomes.
- Charting specific governance, economic, environmental, and social outcomes, goals, and corresponding metrics.
- Recommending the resilience implementation strategy, connecting the management of resilience and sustainability strategic plan to the budget process.
- Mapping out public outreach and campaign plan.

Assigning the Resilience and Sustainability Outcome and Target Champions

A sample process of assigning outcome and target champions and delineating direct and indirect functional responsibilities is provided in Table 7.1, entitled Assigning the Strategic Resilience and Sustainability Plan Outcome and Target Champions. The table represents a resilience and sustainability plan sample, using the Quadruple Bottom Line of governance, economic, environment, and social elements, and layered in the following sequence, Quadruple Bottom Line association > objective > outcome > outcome > outcome > target. It pertains to the significance of assigning direct and indirect champions of resilience and sustainability targets. The template is shared between staff for additional input, changes to the proposed targets, and identifying appropriate stakeholders and outcome and target champions among staff. The numbers and percentages used in the template below are for illustrative purposes.

The proposed plan may be set in the following order: **Quadruple Bottom Line (QBL) association ≫ Objectives ≫ Outcomes ≫ Targets**

The Quadruple Bottom Line provides the overarching elements:

Governance Economic, **Environmental**, and **Social**

The specific elements of the plan are defined through four categories of the Quadruple Bottom Line, starting with Governance and specific targets are categorized under separate goals and outcomes. The four components are associated with resilience and sustainability as an outline of initiatives under the four areas, regardless of what the communities or organizations decided to call it, climate action plan, resilience plan, sustainability plan, climate adaptation and mitigation plan. The logical representation of the objectives and targets are represented in the comprehensive strategic resilience and sustainability plan.

Under Objectives, there are numerous options, tailored to specific needs of the community and organizations:

Under Objectives, each outcome has a broader explanation correlated to targets. The plan is color-coded for ease in identifying spe-

QBL1: GOOD GOVERNANCE

- OBJECTIVE 1: ACCOUNTABLE GOVERNMENT, ACCESSIBLE AND TRANSPARENT SERVICE DELIVERY
- OBJECTIVE 2: EFFECTIVE STAKEHOLDERS ENGAGEMENT
- OBJECTIVE 3: ENSURE AN OPEN AND INCLUSIVE GOVERNMENT
- OBJECTIVE 4: RESILIENT, EFFECTIVE AND EFFICIENT SERVICE DELIVERY
- OBJECTIVE 5: MAINTAIN RESILIENT ASSET MANAGEMENT

QBL 2: RESILIENT ECONOMIC GROWTH, SUSTAINABLE DEVELOPMENT, AND BUSINESS OPPORTUNITIES

- OBJECTIVE 1: CREATE A RESILIENT ENVIRONMENT FOR BUSINESS AND COMMERCE
- OBJECTIVE 2: FACILITATE DIVERSE JOB CREATION AND JOB GROWTH
- OBJECTIVE 3: FOSTER RESILIENT ECONOMIC GROWTH

QBL 3: RESILIENT ENVIRONMENT

- OBJECTIVE 1: RESILIENT ENERGY MANAGEMENT AND CARBON FOOTPRINT REDUCTION
- OBJECTIVE 2: CLIMATE CHANGE RESILIENCE AND READINESS
- OBJECTIVE 3: WASTE MINIMIZATION AND EXPANSION OF RE-USE AND RECYCLING OPPORTUNITIES
- OBJECTIVE 4: PROTECTION AND ENHANCEMENT OF NATURAL SYSTEMS

QBL 4: RESILIENT AND SAFE NEIGHBORHOODS

- **OBJECTIVE 1: PROMOTE RESILIENT QUALITY OF LIFE**
- **OBJECTIVE 2: RESILIENT NEIGHBORHOOD INFRASTRUCTURE**
- **OBJECTIVE 3: INCREASE NEIGHBORHOOD PARTNERSHIPS AND COLLABORATION**

cific objectives, in the sample provided in this chapter, red represents the Governance Quadruple Bottom Line feature association, blue represents the Economic pillar, green is used for Environmental and orange color is used for the Social element of the Quadruple Bottom Line. An illustrative sample of strategic resilience and sustainability plan with targets, in alignment with the Quadruple Bottom Line elements, is provided on the next page.

QBL1: GOOD GOVERNANCE

OBJECTIVE 5: Maintain Resilient Asset Management

OUTCOME 5.1: Implement an integrated lifecycle investment to maintaining the infrastructure and other assets to maximize benefits, ensure resilience, and manage risk and provide satisfactory levels of service to the public in a sustainable and environmentally responsible manner

TARGETS:

1. Adopt an asset management policy and program implementation plan by June 30, 2022.
2. Implement an asset management governance model by June 30, 2022.
3. Establish levels of service measurements consistent with asset management plans by June 30, 2022.
4. Develop a comprehensive five-year capital improvement plan and integrate resilience models by June 30, 2022.

QBL 2: RESILIENT ECONOMIC GROWTH, SUSTAINABLE DEVELOPMENT AND BUSINESS OPPORTUNITIES

OBJECTIVE 3: Foster Resilient Economic Growth and Economy

OUTCOME 1.1: Advance a resilient, entrepreneur-focused economic development strategy leveraging the local government resources, building the internal and external infrastructure required to support the economy, and maintain the economic vitality of the community and the region.

TARGETS:

1. Incentivize $100 million in private investment by June 30, 2030.

QBL 3: RESILIENT ENVIRONMENT

OBJECTIVE 1: Resilient energy management and carbon footprint reduction

OUTCOME 1: Implement initiatives to counteract the effects of greenhouse gas emissions (GHG) to create a resilient community.

TARGETS:

1. Reduce the City's greenhouse gas (GHG) emissions to 15% percent below 2015 levels by 2030.
2. Achieve 100% of energy use from renewable sources from solar, wind, and geothermal by June 30, 2030.
3. Double water reuse and recovery by June 30, 2024.

QBL 4: RESILIENT AND SAFE NEIGHBORHOODS

OBJECTIVE 3: Increase Neighborhood Partnership and Collaboration

OOUTCOME 1: Implement cost-effective, data-driven programs designed for high-risk groups and environments to promote safety, prepare for emergencies, and install and maintain city equipment and systems that ensure a safe and resilient environment for residents and businesses.

TARGETS:

1. Remove graffiti in the city within 24 HOURS from the initial report from neighborhood associations.

Table 7.1 Assigning the strategic resilience and sustainability plan outcome and target champions

TARGET#	TARGET WORDING	Outcome Champion	Comments/Metrics/SUPPORT/REVISIONS/RESEARCH
QBL 1: GOOD GOVERNANCE			
OBJECTIVE 1: ACCOUNTABLE GOVERNMENT AND ACCESSIBLE AND TRANSPARENT SERVICE DELIVERY			
OUTCOME 1: Resilient financial management, and reduced operational costs.			
TARGET 1.1.1.1	Decrease cost as a result of energy efficiency improvements by an additional percentage over FY 2021 results in City facilities by DATE.	Energy Manager	All departments to coordinate with power utilities
OBJECTIVE 2: EFFECTIVE STAKEHOLDERS ENGAGEMENT			
OUTCOME 1: Communicate decision-making process and outcomes in a clear and understandable manner			
TARGET 1.2.1.1	Translate all documents and programs into Bosnian, Spanish, and Arabic.	Analyst	Work with translation agencies.
OBJECTIVE 3: ENSURE AN OPEN AND INCLUSIVE GOVERNMENT			
OUTCOME 1: Ensure services are easily accessible to a diverse customer base through proven best practices and coordination across all service channels			
TARGET 1.3.1.1	Increase the use of online permitting to 100% by 2025.	Planner	City Planning Department
OBJECTIVE 4: RESILIENT, EFFECTIVE AND EFFICIENT SERVICE DELIVERY			
OUTCOME 1: Implement decisions and processes with the most efficient use of resources to serve the needs of the diverse community			
TARGET 1.4.1.1	Respond to 100% of street lighting outages within 24 hours of being reported.	Director	Street Lights & Public Works
1.4.1.2	Increase the City's overall Fire Code inspection completion rate to % by DATE.	Fire Chief	Fire Department & Inspections
QBL 2: RESILIENT ECONOMIC GROWTH, SUSTAINABLE DEVELOPMENT AND BUSINESS OPPORTUNITIES			
OBJECTIVE 1: CREATE A RESILIENT ENVIRONMENT FOR BUSINESS AND COMMERCE			
OUTCOME 1.1: Adopt innovative, entrepreneur-focused economic development strategies that leverage the resources of the city to maintain the economic resilience of the community.			

Tab. 7.1 (continued)

TARGET 2.1.1.1	Attract and retain 25 new businesses in the city before June 30, 2025.	City Manager/ Director	Economic Development Department with the Chamber of Commerce and regional economic development organizations to support the target.
OBJECTIVE 2: FACILITATE DIVERSE JOB CREATION AND JOB GROWTH			
OUTCOME 2.1: Facilitate resilient business development to support job creation using tax incentives and other available economic development tools.			
TARGET 2.2.1.1	90% of jobs created or retained with incentives will be permanent, full-time employment with benefits.	Director	Economic development and business organizations to collaborate and support the target.
TARGET 2.2.1.2	Increase the number of diverse businesses by 30% over FY2020 results by June 30, 2024.	Director	Economic development, diversity officer and local business organizations to collaborate and support the target.
QBL 3: RESILIENT ENVIRONMENT			
OBJECTIVE 1: RESILIENT ENERGY MANAGEMENT AND CARBON FOOTPRINT REDUCTION			
OUTCOME 1: Implement initiatives to counteract the effects of greenhouse gas emissions (GHG) to create a resilient community.			
TARGET 3.1.1.1	Reduce the City's greenhouse gas (GHG) emissions to 15% below 2015 levels by 2030.	Energy Manager	Inventory from 2015 of the total CO2e was estimated to be # of metric tons. TARGET aims to reduce levels to # metric tons CO2e. An alternative TARGET could be --%.... # metric tons CO2e.
TARGET 3.1.1.2	Achieve 100% of energy use from renewable sources from solar, wind, and geothermal by June 30, 2030.	Chief Resilience Officer	Resilience and Sustainability Office
OBJECTIVE 2: CLIMATE CHANGE RESILIENCE AND READINESS			
OUTCOME 1: Integrate climate resilience preparedness into plans to respond to climate change related threats and disasters.			
TARGET 3.2.1.1	Include climate change projections and incorporate climate adaptation planning into capital, operating and maintenance programs by 2022.	Chief Resilience Officer	Resilience and Sustainability Office
TARGET 3.2.1.2	Include climate resilience and vulnerability assessment into emergency preparedness plans by 2022.	Chief Resilience Officer	Resilience and Sustainability Office & Public Safety and Executive Office

Tab. 7.1 (continued)

OBJECTIVE 3: WASTE MINIMIZATION AND EXPANSION OF RE-USE AND RECYCLING OPPORTUNITIES			
OUTCOME 1: Integrate protection and restoration of natural systems into plans to provide ecological services.			
TARGET 3.3.1.1	Reduce the amount of landfill contributions by 20% within 2 years.	Waste Management Director	Solid Waste Management
TARGET 3.3.1.2	Increase tons of waste sent to waste-to-energy facility by 10% each year over the 2020 baseline year.	Energy Director	Energy Management
OBJECTIVE 4: PROTECTION AND ENHANCEMENT OF NATURAL SYSTEMS			
OUTCOME 1: Expand tree planting opportunities and increase plants' bio-diversity in public and private green spaces.			
TARGET 3.4.1.1	Increase tree canopy coverage by 1% per neighborhood by 2025.	Parks Director	Parks Department
TARGET 3.4.1.2	Increase bio-diversity of plants in parks by 1% each year.	Parks Director	Parks Department
QBL 4: RESILIENT AND SAFE NEIGHBORHOODS			
OBJECTIVE 1: PROMOTE RESILIENT QUALITY OF LIFE			
OUTCOME 1: Promote quality design and construction consistent with neighborhood character to encourages efficient land use, green building design, sustainable mobility, resilience, safety, walkability.			
TARGET 4.1.1.1	Decrease the number of vacant lots or brownfields by 10% per neighborhood increasing by 2030. .	Community Development Director	Metric: % of brownfields per neighborhood & % of projects preserving historic buildings. Alternative metrics: % of land designated by city to preserve open space % of land designated as historic preservation
TARGET 4.1.2.1	Increase the number of completed lots to preserve historic buildings and repurpose vacant lots and brownfields by 10% by 2030.	Planning Director	Planning department
OBJECTIVE 2: RESILIENT NEIGHBORHOOD INFRASTRUCTURE			
OUTCOME 2: Promote resilient neighborhood infrastructure withhousing, city streets, and sidewalks, available trees, green space, access to parks and recreation amenities as important elements of any neighborhood.			
TARGET 4.2.2.1	Increase the number of affordable housing units by 100 by June 30, 2024.	Community Development Director	COMMUNITY DEVELOPMENT

Resilient Emergency Management, Disaster and Pandemic Preparedness, and Climate Readiness

Climate change represents an unprecedented set of challenges to emergency management and planning, disaster readiness, and preparedness for municipalities. Borchers et al. (2020), Reid et al. (2016), Richardson et al. (2012), Smith et al. (2014), and Youssouf (2014) documented the health impacts and social costs associated with wildfires and heatwaves, poor air quality, further exacerbated by climate change and global warming.

The recent COVID-19 pandemic has put an additional strain on all resources in communities and organizations in all sectors and put a spotlight on public health risks from the global pandemic. Communities continue to experience the effects of the extended heatwave events, fires, COVID-19 and droughts with long-term consequences.

To appropriately and timely respond to disasters and generate resilient emergency management, cities, counties, and states emergency action guidelines, disaster preparedness, and hazard mitigation plans are amended to include additional considerations and planning procedures incorporating the climate change information and data.

The following checklist aids the public health and communication service sections in the emergency preparedness plans:

- Coordinate with the nonprofit organizations to establish cooling shelters that provide air conditioning and relief from extreme heat events and fires.
- Distribute public statements to inform residents of ways to avoid heat-related illness and injury, encouraging them to check on vulnerable neighbors and family members and providing them with the shelter locations information.
- Issue public statement to encourage lowering emissions during prolonged heat events to improve air quality.
- Organize and coordinate with power utilities to ensure timely response in the case of prolonged power outages.
- Ensure all pandemic responses are coordinated with public health agencies.

Beyond amendments to the governments' emergency preparedness guidelines and plans, the list of recommendations and supporting documents are included in the Public Health Services section of the emergency and disaster plans. The items are incorporated into a comprehensive report on emergency preparedness in response to extreme weather threats and the changing climate.

Resilient Emergency Planning Recommendations:

- Inventory available communication methods and the crucial emergency personnel.
- Catalog available communication methods with the community concerning emergencies and available resources and solutions, including landlines and other means of communication in case of loss of landlines and cellular reception.
- Assess the ability to identify specific portions of the city and population, low income and vulnerable population, including elderly, school children, businesses, and community aid organizations.
- Evaluate the ability to disseminate emergency information to specific segments of the population.
- Create educational material on the likelihood of an increased number of emergency events.
- Provide the educational content on the availability of recourses for a particular emergency event: including evacuation routes, cooling center locations, water pollution threats and potential treatments, avoiding injury or health issues during high heat and poor air quality events.
- Include stakeholders from various sectors to ensure all concerns are addressed comprehensively.
- In the case of a pandemic, it is critical to coordinate efforts with the federal, state, and local public health agencies.

Severe Heatwaves or Fires Events Resilient Planning Recommendations:

- Use satellite imagery, census data, and GIS to identify vulnerable populations.
- Use satellite imagery and GIS to determine locations where the heat island effect will increase the severity of heatwave events or fires.
- Reduce carbon emissions during high heat events to improve the local air quality.
- Create an automated system to alert and check on at-risk citizens.

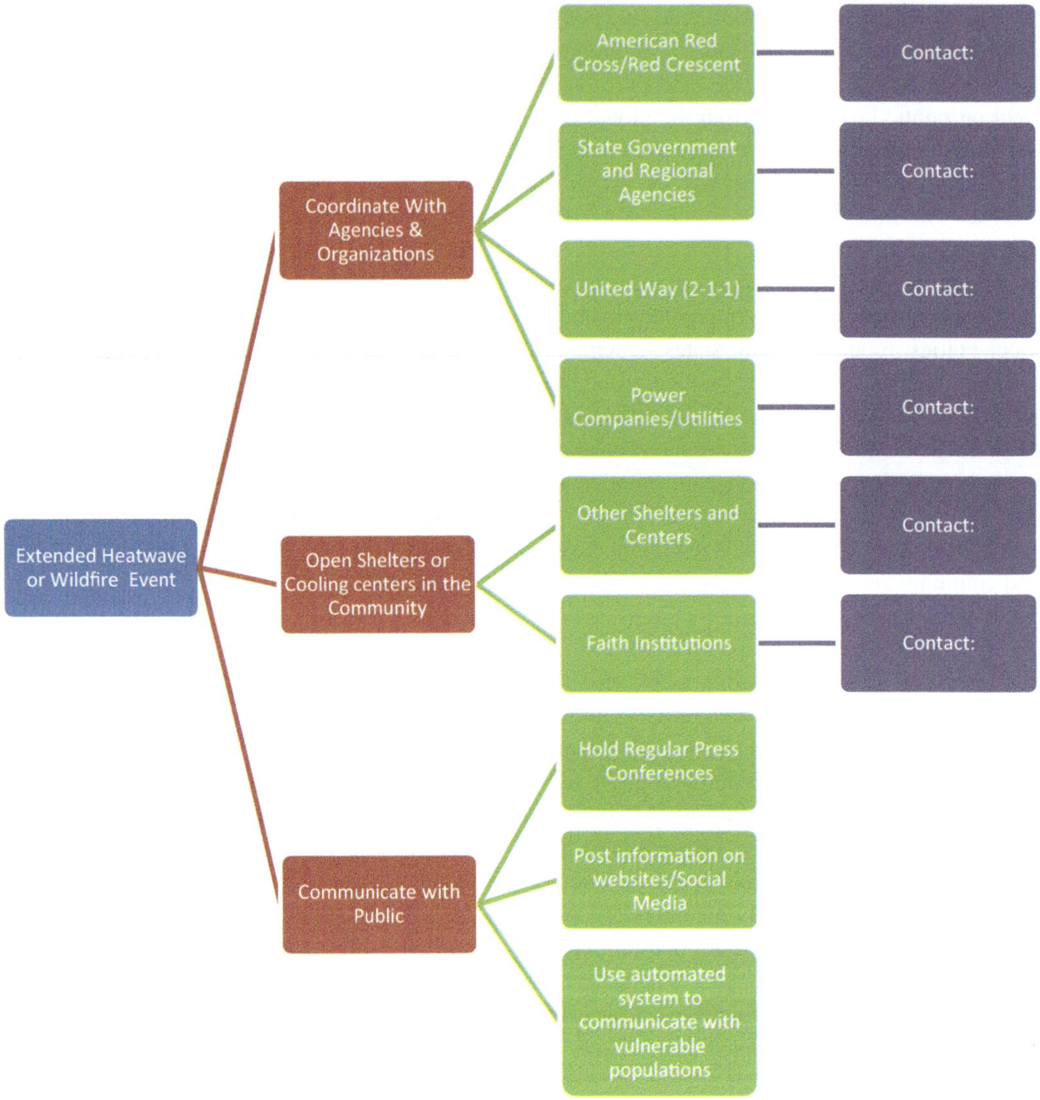

Fig. 7.1 Flowchart—emergency preparedness for extended heatwave or wildfires events

- Open the cooling centers or centers for evacuated populations and collaborate with local nonprofit organizations to make temporary shelters.

In the case of protracted heat events or wildfires, the flowchart represented by Fig. 7.1 features a recommended community-wide response, from coordinating with the relevant community organizations, utilities, and state, regional, and federal agencies, communicating with the public, and opening of the shelters (Fig. 7.1).

Pandemic Resilient Planning Recommendations:

- Provide easy access and infrastructure to testing and vaccination for residents.
- Ensure daily data on new cases and hospitalizations, mortality rates.
- Coordinate shutdowns and shelter in place with the state and regional agencies.
- Strategize mask policies and other mandates and requirements across all levels of governments.

Summary

Countless resources and tools are available at no or at a relatively small cost to the organizations and communities to develop robust strategic resilience and sustainability plan. The availability and ease of access to resources make it a compelling reason for governments to pursue resilience and sustainability initiatives in their plans. Likewise, with the supportive elected and appointed leadership, institutionalized resilience and sustainability strategies, and dedicated staff, the use of available resources and tools results in concrete strategies and outcomes for the organization. Moreover, the available resources and tools are only relevant if the appropriate internal and external stakeholders are identified, environmental analysis is conducted, and measurable targets are created using the steps described in the previous chapters.

Finally, the tools and resources add value with dedicated staff to plan and implement strategic resilience and sustainability initiatives. Institutionalizing strategic resilience and sustainability planning assists organizations and communities to better prepare for the inevitable adverse effects of climate change and be ready for events such as the current COVID-19 pandemic to be more resilient to the unpredictable nature of disasters.

Outcomes, Discussions, and Further Considerations

- Discuss the resources that local government administrators would require for resilience and sustainability strategic planning.
- Assess how the communities and organizations propose to measure carbon footprint and energy productions.
- Scrutinize the available measures utilized by organizations and communities.
- Evaluate the available staffing and the requirements for implementing a strategic resilience and sustainability plan.
- Analyze the objectives and targets, including reducing energy consumption, renewable energy, and greenhouse gas emission targets for communities and organizations.
- Discuss how to implement climate resilience, pandemic responses, disaster readiness, and resilient emergency management.

References

Alibašić H (2018) Role of corporations in addressing climate change. In: Farazmand A (ed) Global encyclopedia of public administration, public policy, and governance. Springer International Publishing AG, Cham. https://doi.org/10.1007/978-3-319-31816-5_3429-1

Alibašić H (2020) Sustainable procurement. In: Farazmand A (ed) Global encyclopedia of public administration, public policy, and governance. Springer International AG, Cham. https://doi.org/10.1007/978-3-319-31816-5_3427-1

Architecture 2030 (2018a) About us. https://architecture2030.org/about/

Architecture 2030 (2018b) 2030 challenge. https://architecture2030.org/2030_challenges/2030-challenge/

Borchers AN, Bowman DMJS, Palmer AJ, Johnston FH (2020) Climate change, wildfires, heatwaves and health impacts in Australia. In: Akhtar R (ed) Extreme weather events and human health. Springer, Cham. https://doi.org/10.1007/978-3-030-23773-8_8

Brown ME, Antle JM, Backlund P, Carr ER, Easterling WE, Walsh MK, Ammann C, Attavanich W, Barrett CB, Bellemare MF, Dancheck V, Funk C, Grace K, Ingram JSI, Jiang H, Maletta H, Mata T, Murray A, Ngugi M, Ojima D, O'Neill B, Tebaldi C (2015) Climate change, global food security, and the U.S. Food System. 146 pages. http://www.usda.gov/oce/climate_change/FoodSecurity2015Assessment/FullAssessment.pdf

City of Houston (2017) EPA's climate change is real website. http://www.greenhoustontx.gov/climate-change-is-real.html

City of Milwaukee (2017) Climate change is real. http://city.milwaukee.gov/sustainability/About-Us/Climate-Change.htm#.WmzTpZM-fys

Davis J, Alibašić H, Norris S (2018) Corporate leadership in sustainability: a green ranking performance-based approach to understanding corporate social responsibility (CSR) and positive marketing impact. Creighton J Interdiscip Leadersh. 4(1):3–6

Global Covenant of Mayors for Climate and Energy (2021) Global covenant of mayors for climate and energy. https://www.globalcovenantofmayors.org/

Great Lakes Climate resilience (2013) Great lakes climate resilience planning guide. http://greatlakesresilience.org/

Great Lakes Saint Lawrence Cities Initiative (GLSLCI) (2021a) About us. https://glslcities.org/about/
Great Lakes Saint Lawrence Cities Initiative (GLSLCI) (2021b) Initiatives. https://glslcities.org/initiatives/
Great Lakes Saint Lawrence Cities Initiative (GLSLCI) (2021c) Climate change adaptation. https://glslcities.org/initiatives/municipal-climate-adaptation/
Hardcastle JL (2016a) How PepsiCo saved $80 million by cutting water use 26%. Environmental + Energy Leader. https://www.environmentalleader.com/2016/09/how-pepsico-saved-80-million-by-cutting-water-use-26/#ixzz4NMniOZuF
Hardcastle JL (2016b) How Ford, PepsiCo Plan to slash manufacturing water use. Environmental + Energy Leader. https://www.environmentalleader.com/2016/10/how-ford-pepsico-plan-to-slash-manufacturing-water-use/
Ibn Khaldun AAbM (1406, 1969) The Muqaddimah: an introduction to history; in three volumes. (Franz Rosenthal, Trans.) Princeton University Press, New Jersey. Abridged edition
ICLEI – Local Governments for Sustainability USA (ICLEI USA) (2021a) Who we are. https://icleiusa.org/about/who-we-are/
ICLEI – Local Governments for Sustainability USA (ICLEI USA) (2021b) Tools. https://icleiusa.org/clearpath/
Intergovernmental Panel On Climate Change (IPCC) (2021a) Climate change 2021: the physical science basis. In: Masson-Delmotte V, Zhai P, Pirani A, Connors SL, Péan C, Berger S, Caud N, Chen Y, Goldfarb L, Gomis MI, Huang M, Leitzell K, Lonnoy E, Matthews JBR, Maycock TK, Waterfield T, Yelekçi O, Yu R, Zhou B (eds) Contribution of working group I to the sixth assessment report of the intergovernmental panel on climate change. Cambridge University Press, Cambridge
Intergovernmental Panel On Climate Change (IPCC) (2021b) IPCC working group I (WGI): sixth assessment report. Interactive Atlas. https://interactive-atlas.ipcc.ch/
Intergovernmental Panel On Climate Change (IPCC) (2021c) IPCC AR6 sea level projection tool. https://sealevel.nasa.gov/data_tools/17
Melillo JM, Richmond TC, Yohe GW (2014) Climate change impacts in the United States: the third national climate assessment. U.S. Global Change Research Program, 841 pp. https://doi.org/10.7930/J0Z31WJ2.
Murray T (2019) The businesses that are—and are not—leading on climate change. Environmental Defense Fund. Forbes. https://www.forbes.com/sites/edfenergyexchange/2019/11/08/the-businesses-that-are%2D%2Dand-are-not%2D%2Dleading-on-climate-change/?sh=5d9707447aa1
National League of Cities (NLC) (2017) National league of cities launches local climate solutions engagement program. http://www.nlc.org/article/national-league-of-cities-launches-local-climate-solutions-engagement-program
National League of Cities (NLC) (2021) Sustainability and resilience. https://www.nlc.org/initiative/sustainability-and-resilience/
National Oceanic and Atmospheric Administration (NOAA) (2017) Digital coasts tools. https://coast.noaa.gov/digitalcoast/tools/
National Oceanic and Atmospheric Administration (NOAA) (2018) Great lakes water level. NOAA—Great Lakes Environmental Research Laboratory. https://www.glerl.noaa.gov/data/wlevels/levels.html
National Oceanic and Atmospheric Administration (NOAA) (n.d.) U.S. climate resilience toolkit. https://toolkit.climate.gov/
National Oceanic and Atmospheric Administration (NOAA) Climate.gov (n.d.) data snapshot. https://www.climate.gov/maps-data/data-snapshots/start
Northern Gulf of Mexico Sentinel Site Cooperative (2021a) About the Northern Gulf of Mexico Sentinel Site Cooperative. http://masgc.org/northern-gulf-of-mexico-sentinel-site-co/about
Northern Gulf of Mexico Sentinel Site Cooperative (2021b) About Gulf TREE. http://masgc.org/northern-gulf-of-mexico-sentinel-site-co/tree
Northern Gulf of Mexico Sentinel Site Cooperative (2022). About the Northern Gulf of Mexico Sentinel Site Cooperative. https://masgc.org/northern-gulf-of-mexico-sentinel-site-co
Reid CE, Brauer M, Johnston FH, Jerrett M, Balmes JR, Elliott CT (2016) Critical review of health impacts of wildfire smoke exposure. Environ Health Perspect 124(9):1334–1343. https://doi.org/10.1289/ehp.1409277
Resilient Cities Network (2021) Our story. https://resilientcitiesnetwork.org/our-story/
Resilient Michigan (2017) Planning for community resilience in Michigan: a comprehensive handbook. http://www.resilientmichigan.org/downloads/michigan_resiliency_handbook_web.pdf
Resilient Michigan (n.d.) Resilient Michigan Communities. Retrieved from http://www.resilientmichigan.org/about.asp
Richardson LA, Champ PA, Loomis JB (2012) The hidden cost of wildfires: economic valuation of health effects of wildfire smoke exposure in Southern California. J For Econ 18(1):14–35. https://doi.org/10.1016/j.jfe.2011.05.002
Smith KR, Woodward A, Campbell-Lendrum D, Chadee D, Honda Y, Liu Q, et al (2014) Human health: impacts, adaptation, and co-benefits. In: Field CB, Barros VR, Dokken DJ, Mach KJ, Ma (ed) Climate change 2014: impacts, adaptation, and vulnerability. Part A: global and sectoral aspects. Contribution of working group II to the fifth assessment report of the intergovernmental panel on climate change. https://doi.org/10.1017/CBO9781107415379.016.
The Alliance for a Sustainable Future (2020) Mayors leading the way on climate how cities large and small are taking action. https://www.c2es.org/site/assets/uploads/2020/01/Mayors-Leading-the-Way-Volume-III.pdf

The Compact of Mayors (n.d.) The compact of mayors goals and objectives. https://www.c40.org/researches/compact-of-mayors

The National Aeronautics and Space Administration (NASA) (2021a) Sustainability and government resources. https://climate.nasa.gov/solutions/resources/

The National Aeronautics and Space Administration (NASA) (2021b) Mitigation and adaptation. https://climate.nasa.gov/solutions/adaptation-mitigation/

The National Aeronautics and Space Administration (NASA) (2021c) Earth science in action. https://climate.nasa.gov/solutions/earth-science-in-action/?page=0&per_page=40&order=publish_date+desc%2C+created_at+desc&search=&category=147

The Nature Conservancy (2021a) Center for Sustainability Science. https://www.nature.org/en-us/about-us/who-we-are/our-science/center-for-sustainability-science/

The Nature Conservancy (2021b) Our priorities: tackling climate change. https://www.nature.org/en-us/what-we-do/our-priorities/tackle-climate-change/

The Rockefeller Foundation (2021) 100 resilient cities. https://www.rockefellerfoundation.org/100-resilient-cities/

The United States Conference of Mayors (USCM) (2021) Mayors climate protection center. https://www.usmayors.org/programs/mayors-climate-protection-center/

The United States Environmental Protection Agency (US EPA) (2020) Pollution prevention tools and calculators. https://www.epa.gov/p2/pollution-prevention-tools-and-calculators

The United States Environmental Protection Agency (US EPA) (2021a) Pollution prevention program. https://www.epa.gov/p2

The United States Environmental Protection Agency (US EPA) (2021b) Climate change. https://www.epa.gov/climate-change

The United States Environmental Protection Agency (US EPA) (2021c) Comprehensive procurement guideline (CPG) program. https://www.epa.gov/smm/comprehensive-procurement-guideline-cpg-program

The United States Global Change Research Program (USGCRP) (n.d.) Fifth National Climate Assessment. Retrieved from https://www.globalchange.gov/nca5

The United States Global Change Research Program (USGCRP) (2016) In: Crimmins A, Balbus J, Gamble JL, Beard CB, Bell JE, Dodgen D, Eisen RJ, Fann N, Hawkins MD, Herring SC, Jantarasami L, Mills DM, Saha S, Sarofim MC, Trtanj J, Ziska L (eds) The impacts of climate change on human health in the United States: a scientific assessment. U.S. Global Change Research Program, Washington, DC. , 312 pp. https://doi.org/10.7930/J0R49NQX

The United States Global Change Research Program (USGCRP) (2017) In: Wuebbles DJ, Fahey DW, Hibbard KA, Dokken DJ, Stewart BC, Maycock TK (eds) Climate science special report: fourth national climate assessment, volume I. U.S. Global Change Research Program, Washington, DC. , 470 pp. https://science2017.globalchange.gov/. https://doi.org/10.7930/J0J964J6.

The United States Global Change Research program (USGCRP) (2018a) In: Cavallaro N, Shrestha G, Birdsey R, Mayes MA, Najjar RG, Reed SC, Romero-Lankao P, Zhu Z (eds) Second state of the carbon cycle report (SOCCR2): a sustained assessment report. U.S. Global Change Research Program, Washington, DC. ,878 pp. https://carbon2018.globalchange.gov/. https://doi.org/10.7930/SOCCR2.2018

The United States Global Change Research program (USGCRP) (2018b) In: Reidmiller DR, Avery CW, Easterling DR, Kunkel KE, Lewis KLM, Maycock TK, Stewart BC (eds) Impacts, risks, and adaptation in the United States: fourth national climate assessment, volume II. U.S. Global Change Research Program, Washington, DC. , 1515 pp. https://doi.org/10.7930/NCA4.2018

The United States Global Change Research Program (USGCRP) (n.d.) Fifth national climate assessment. https://www.globalchange.gov/nca5

The White House (2014) President Barack Obama. President's state, local, and tribal leaders task force on climate preparedness and resilience. Recommendations to the President. Retrieved from https://obamawhitehouse.archives.gov/sites/default/files/docs/task_force_report_0.pdf

The White House (2021a) Executive order on protecting public health and the environment and restoring science to tackle the climate crisis. https://www.whitehouse.gov/briefing-room/presidential-actions/2021/01/20/executive-order-protecting-public-health-and-environment-and-restoring-science-to-tackle-climate-crisis/

The White House (2021b) Paris climate agreement. https://www.whitehouse.gov/briefing-room/statements-releases/2021/01/20/paris-climate-agreement/

The White House (2021c) Executive order on tackling the climate crisis at home and abroad. https://www.whitehouse.gov/briefing-room/presidential-actions/2021/01/27/executive-order-on-tackling-the-climate-crisis-at-home-and-abroad/

U.S. Fish & Wildlife Services (2021a) Conservation in a changing climate. https://www.fws.gov/home/climatechange/

U.S. Fish & Wildlife Services (2021b) Climate change. https://www.fws.gov/refuges/wildlife-conservation/climate-change.html

United Nations Framework Convention on Climate Change (UNFCCC) (2021a) About the Secretariat. https://unfccc.int/

United Nations Framework Convention on Climate Change (UNFCCC) (2021b) About the Secretariat. https://unfccc.int/about-us/about-the-secretariat

United Nations Framework Convention on Climate Change (UNFCCC) (2021c) Annual report 2021. https://unfccc.int/sites/default/files/resource/UNFCCC_Annual_Report_2020.pdf

United Nations Framework Convention on Climate Change (UNFCCC) (2021d) Annual report 2021. https://unfccc.int/about-us/partnerships/unfccc-partners
University of Wisconsin-Madison (2016) Optimizing CIP to save money and reduce waste. https://engrshwec.wiscweb.wisc.edu/wp-content/uploads/sites/711/2015/08/CIP-Fact-Sheet-fin-3-11-16.pdf
Urban Sustainability Directors Network (USDN) (2021) About USDN. https://www.usdn.org/about.html
Xu Z, Mim NK, Franchetti M, Kumar A (2016) A facility lighting comparison based on energy savings and efficiency, pollution prevention and life cycle assessment. Macrothink Institute, Las Vegas. https://doi.org/10.5296/emsd.v5i2.9801
Youssouf H, Liousse C, Roblou L, Assamoi EM, Salonen RO, Maesano C et al (2014) Non-accidental health impacts of wildfire smoke. Int J Environ Res Public Health 11(11):11772–11804. https://doi.org/10.3390/ijerph111111772

8 Envisaging the Future of Strategic Resilience and Sustainability Planning

"The complete problem of climatic change entails several distinct sub-problems. First, there is the observational task of establishing that changes of climate actually have occurred-by no means a trivial undertaking-and of determining the nature and extent of these changes. At the other extreme, there is the theoretical task of determining just what changes in climate would take place as a result of specified hypothetical causes. An intermediate problem is that of identifying the principal cause or causes of those changes in climate which have actually happened."

Lorenz (1970, p. 325)

Overview

Chapter eight envisions the future of strategic resilience and sustainability planning in communities and organizations. The chapter evaluates the complexities of the modern dynamics of the interconnected systems in organizations and communities and recommends how city and county administrators must consider these challenges in planning methods and strategies. The chapter examines the current concerns and future threats in communities and organizations surrounding pandemics, disaster resilience, climate change, and emergency management and preparedness. It presents the private sector perspectives on resilience and sustainability. Planning by organizations includes resilience and sustainability components. Beyond climate change, cities and counties, communities and organizations, public and private sectors will face other challenges and threats in the future, as seen during the COVID-19 pandemic. Embedding resilience and sustainability into strategic initiatives and plans maximizes the outcomes for organizations and communities.

Keywords

Adaptability · Innovation · Responsiveness
Good governance · Climate resilience
Climate preparedness · Transformation
Interconnected systems · Partnership
Leadership · Reporting · Cross-sectoral
COVID-19 pandemic · System-wide planning

Key Questions

The eighth and final chapter of this book summarizes previous sections and discusses the following underlying premises and topics.

- What is the future of strategic resilience and sustainability planning in communities and organizations?
- Beyond climate change what challenges and barriers will communities and organizations face in the future?
- How do organizations increase investing in resilience and sustainability initiatives and projects?
- How can resilience be embedded and institutionalized within organizations?

H. Alibašić, *Strategic Resilience and Sustainability Planning*, Sustainable Development Goals Series, https://doi.org/10.1007/978-3-030-91159-1_8

Introduction

Strategic resilience and sustainability planning analyzes the complexities and dynamics of interconnected systems. Administrators, managers, political leaders, and policy-makers around the globe seek the holistic, systematic, and pioneering solutions for service and production delivery. Cities and counties' leaders maintain and enhance their assets, support the environment and human resources, and invest in the modernization of operations to provide cost-effective and resilient services to residents and businesses communities. Modern administrators design municipalities and the support systems using resilience and sustainability planning for cities and communities to counter emergent and climate change threats. Leading complex systems require a systematic approach to problem-solving, coupled with significant leadership, partnership, and innovation (Alibašić 2018a, b, c).

The Resilience and Sustainability Planning in Contemporary Urban Settings

The world energy use and carbon emissions resulting from energy production using coal and natural gas is staggering in size. Urban cities are major contributors to greenhouse gas emissions and pollution. While covering less than 2% of the planet, cities are directly responsible for over 60% of the world's greenhouse gas emissions and use almost 80% of all the energy produced (Satterthwaite 2008; United Nations n.d.; UNEP 2020).

In the recent analysis of over 160 cities around the world and their greenhouse gas emissions, Wiedmann et al. (2020) and Wei et al. (2021) concluded that 25 cities contribute over 50% of the greenhouse gas emissions globally. United Nations Department of Economic and Social Affairs Population Division (2018) reported that 55% of the world population were living in urban settings, with the projections showing almost 70% of the world population to live in cities by 2050. The United Nations (2015) sustainable development goals call for strengthening resilience and adaptive capacity to counter threats of climate change and related hazards and natural disasters, integrating climate change measures into strategies and improving education and awareness on climate change. With the increased threats of climate change and, more recently of the global pandemic, cities are faced with the most daunting of tasks, to provide services and plan for short and long-term effects and consequences of disasters. Since the pandemic began, in the United States alone, over 880,000 people have died from COVID-19 (Johns Hopkins University & Medicine 2021, 2022). The threats of the spread of the virus in urban settings are exacerbated by the use of mass transportation and population density. City and county administrators, policy-makers, and government officials have a duty and hold a responsibility to prepare and respond to crises robustly and effectively.

Cities around the world increased initiatives to prepare for climate change-related impacts and other threats. With organizational goals and objectives to increase the resilience of communities, organizations face various obstacles beyond financial challenges. Administrators and managers readjust the planning in organizations and communities, emphasizing building resilience. Municipalities aim to reduce greenhouse gas emissions, mitigate climate change, and adapt to new climate uncertainties to increase climate resilience in cities.

A robust inventory of energy use, and carbon footprint and greenhouse gas emissions is crucial in executing the effective mitigation measures and deploying them in organizations and communities. Cities Alliance (n.d.) is a global network and partnership aimed at supporting cities climate resilience efforts. The carbonn center (2021) offers a unified reporting platform for cities, towns, villages, and communities to inventory their climate mitigation and adaptation efforts with a map of participating cities and their climate resilience initiatives. Moreover, the Global Protocol for Community-Scale Greenhouse Gas Emission Inventories (GPC) developed by the World Resource Institute, C40, and ICLEI provides a solid framework and methodologies to calculate, inventory, and report the greenhouse gas emissions in cities and communities (Greenhouse Gas Protocol n.d., p. 10). The Greenhouse Gas Protocol (n.d.) divides greenhouse gas emissions into six distinct categories:

- Stationary energy from residential buildings; commercial and institutional buildings and facilities; manufacturing industries and construction; energy industries, agriculture, forestry and fishing activities; nonspecified sources; fugitive emissions from mining, processing, storage, and transportation of coal; and fugitive emissions from oil and natural gas systems.
- Transportation from on-road, railways, waterborne navigation, aviation, and off-road.
- Waste from solid waste disposal, biological treatment of waste, incineration and open burning, and wastewater treatment and discharge.
- Industrial processes and product use (IPPU).
- Agriculture, forestry, and other land use (AFOLU).
- Other emissions occurring outside the geographic boundary as a result of city activities.

In inventorying greenhouse gas emissions, incorporating climate preparedness with vulnerabilities assessment into strategic plans and emergency action guidelines, cities prepare and plan for negative impacts from extreme weather events and climate change. The resilience and sustainability planning takes into consideration all the elements of the cities' systems for service provisions, including power, water and sewer infrastructure, solid waste management, buildings, health, and public safety, and incorporates them into a single resilience and sustainability plan. The integrated, multi-modal, multi-system approach is shown in the graph in Fig. 8.1, featuring elements and processes of climate mitigation and adaptation that are interspersed and connected. Figure 8.1 represents the Quadruple Bottom Line approach to resilience and sustainability planning with the governance, economic, environmental, and social elements integrated into the strategic plan (Alibašić 2018d).

The crucial components of the strategic planning process are community engagements, partnership, and collaborative opportunities, reporting mechanism, and improvements. Administrators and managers in the organizational and leadership capacity adopt the strategies, recognizing key stressors. Advancing long-term situational analysis and incorporating vulnerability assessment are instrumental in guiding cities, communities, and organizations to avoid the adverse consequences on organizations and communities in case of natural disasters and crises.

Transitioning from Sustainable to Resilient Development

The Brundtland Commission also recognized as the World Commission on Environment and Development (WCED) defined sustainable development through the lens of future generations and their needs (United Nations General Assembly 1987). In principle, sustainable development is an attempt by organizations and individuals to minimize the negative environmental effects from their activities while pursuing positive societal outputs, improving governance, and maximizing the economic gains. Resilient development is characterized as an expansion of sustainable development in recognition of existential threats of climate change, pandemics, and other unforeseen circumstances.

Strategic resilience and sustainability planning affords organizations to use a multidimensional, cross-sectoral, and system-wide approach for the betterment and operational efficiency of organizations around the globe, from cities to multinational corporations. Organizational leaders implement resilience and sustainability strategies to address governance, economic, environmental, and social factors resulting from climate change and natural disasters.

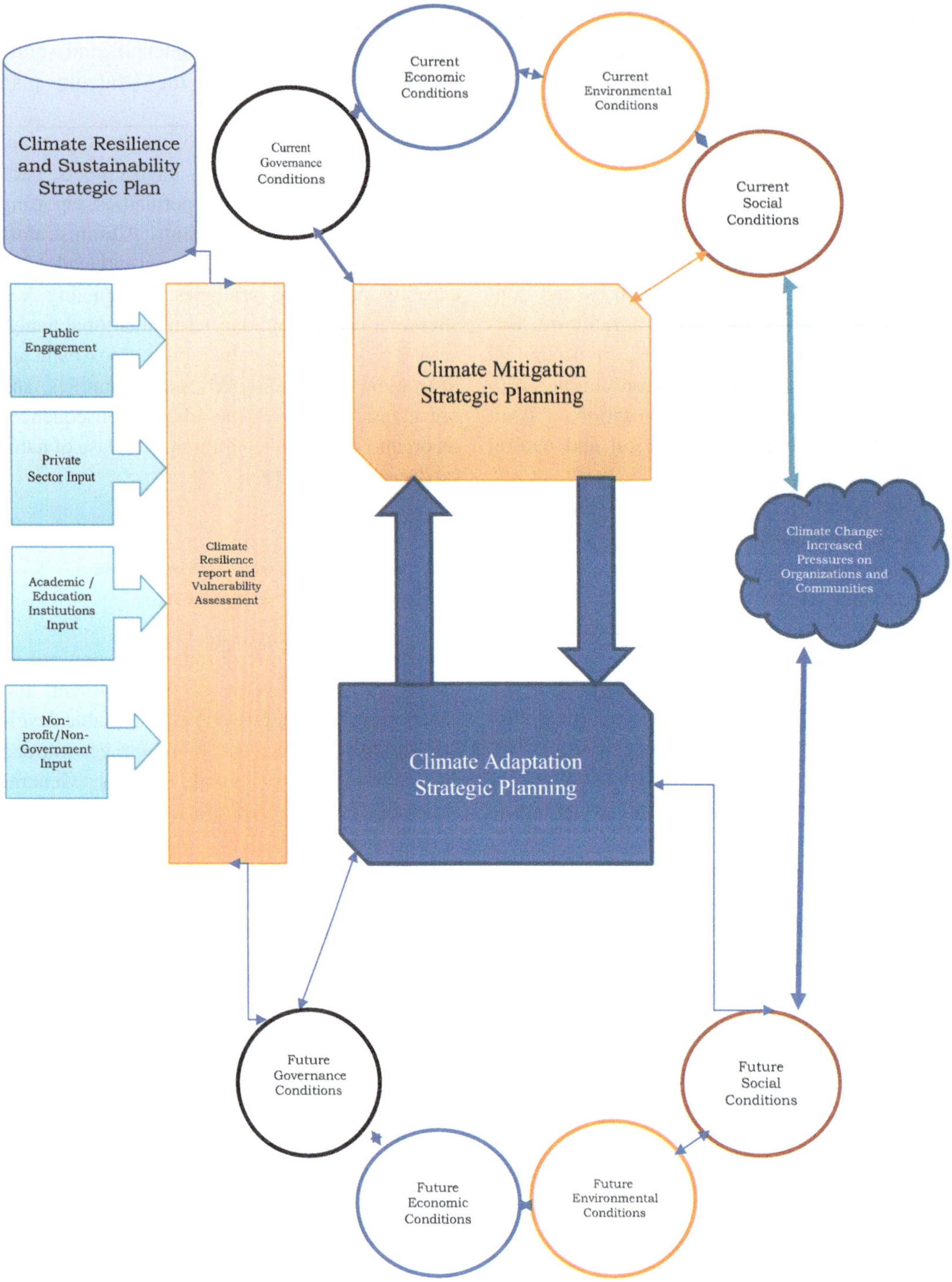

Fig. 8.1 Climate resilience and sustainability strategic plan: climate mitigation and adaptation strategies using the quadruple bottom line

The Organizational Culture and the Imminence of Resilience and Sustainability Planning

The primary indicators of thriving resilience and sustainability projects, programs, and initiatives are the abilities of communities and organizations to transform and adapt to the changing governance, economic, environmental, and societal conditions surrounding them. In linking the organizational culture to people's personalities, Osborne and Plastrik (1997) posited the organizational culture changes are difficult. Weick (1993, 1995) analyzed the organizations and improvisation and Austin (1997) developed a method to study controversial social changes in organizations.

In the research on organizational changes and development, Weick and Quinn (1999) concluded the "change starts with failures to adapt and that change never starts because it never stops" (p. 381). Furthermore, "classic machine bureaucracies, with their reporting structures too rigid to adapt to faster-paced change, have to be unfrozen to be improved" (Weick and Quinn 1999, p. 381). The changes in organizational culture are necessary from within and from outside and are critical for implementing resilience and sustainability in organizations and communities. Both the private and the public sector organizations deploy resilience and sustainability activities to meet the customers' demands. Some cities have adopted more aggressive planning to include resilience and sustainability strategies (Washington County n.d.; City of Golden Valley 2021, n.d.).

However, public organizations and governments have different priorities, and the political flavors of elected officials influence the decision-making process, often clouded by political aspirations and favors. In studying and analyzing municipal climate action plans in 885 EU cities, Reckien et al. (2018) concluded, "city size, international climate networks and national regulation are influential parameters in driving the development of Local Climate Plans" (p. 218). Furthermore, in their study of 327 cities in the EU, Salvia et al. (2021) found that among other factors reaching the carbon neutrality goals depended on their size and combination of the mitigation and adaptation plans into a single strategy. Residents and businesses in cities seek reliable and robust services. Municipal officials consider the complex nature of communities and design plans to enhance the resilience of communities to security threats, natural catastrophes, emergencies, pandemics, and climate change.

Communities and organizations are complex systems pursuing practical resilience, sustainability, and related programs and policies. Without a system-wide approach, even the most pragmatic, robust, and resilient solutions are lost in the administrative and collective nuances. The strategic resilience and sustainability planning method is advantageous to organizations and communities in managing complexities, advancing adaptability and innovation, encouraging good governance and answerability, promoting public engagement, collaboration, and partnerships for more transparency. In brief, strategic resilience and sustainability planning affords organizations to:

Adapt to Complexities and the Modern Dynamics of the Interconnected Systems

Administrators and managers pursue holistic inventions and innovate to maintain and enhance the organizational assets, advance environmental improvements, and support human resources. The complex nature of communities and organizations require the integrated approach to transportation, energy, water systems, solid waste, mobility, infrastructure, diversity, and other elements of resilient communities to address and to enhance resilience initiatives from security threats, extreme weather, fiscal pressures, emergencies, and climate change. Managing complex systems requires a systematic problem-solving approach and methodology.

Advance Adaptability and Innovation in Organizations

Resilience and sustainability planning offers an opportunity for organizations to adapt to changing circumstances surrounding organizations. Administrators and managers integrate a holistic approach to innovating in service delivery. Furthermore, contemporary organizations design the operational systems to adapt to a changing, dynamic environment for maximum resilience to withstand pressures and react to threats.

Encourage Good Governance and Answerability

Changes in the external and internal organizational factors and societal priorities lead to increased responsiveness, identified priorities, and shared responsibilities. Moreover, adjusted approaches are undertaken by organizational leadership in responding to demands and engaging on issues. Cities and other communities invest in long-term resilient initiatives to improve the overall effectiveness and efficiency of the organizational service delivery with a robust return on investment. As an illustration, cities investing in sustainable energy see a social, environmental, and economic return on energy efficiency and renewable energy investments. In return, those savings are reinvested into service provision and improve overall governance and increase accountability and answerability.

Promote Public Engagement, Collaboration, and Partnerships for More Transparency

The key to a robust strategic resilience and sustainability plan is a participatory stakeholder engagement, meaningful input, and rapport. The more inclusive and diverse the local governments are with intentional public participation in the planning processes, the more transparent and answerable organizations are with improved governance and positive outcomes. Municipalities have been partnering with universities, non-governmental agencies, the private sector, and non-profit agencies in delivering the resilience and sustainability outcomes (Alibašić 2017 & 2020; Alibašić and Crawley 2018, 2020).

Summary

The research in this book represents theoretical postulates and practical frameworks of prosperous and durable strategic resilience and sustainability planning. The research relies on a vast volume of literature in climate change studies, resilience theories, sustainability, and climate resilience. It is a practical guide with elements necessary for developing and applying resilience and sustainability plans, actions, and programs. An in-depth review of existing programs, resilience plans, sustainability, and climate action initiatives reveals the current best practices in the field.

The eight chapters provide the detailed outline to initiating, developing, and implementing a robust, strategic resilience and sustainability plan, with the undermentioned components defined:

- Delineating and explaining differences and similarities between resilience and sustainability planning and strategic plan.
- Charting out the resilience and sustainability strategies for organizations and communities.
- Identifying the internal and external stakeholders, the level of organizational and community engagement, partnerships, and outcome champions.
- Measuring, tracking, observing, scrutinizing, and reporting resilience and sustainability targets and outcomes using the Quadruple Bottom Line (QBL) methodology.
- Implementing the resilience and sustainability strategies, illuminating programs and initiatives to accomplish resilient and sustainable communities.
- Examining the intersection of resilience, sustainability, and climate change and the corresponding outcomes of the planning process.

- Emphasizing available tools, methods, and resources for strategic resilience and sustainability planning.
- Recognizing the importance of the private sector in delivering resilience and sustainability outcomes.
- Considering the risks associated in contemporary cities and communities from climate change to pandemics.

Resilience and sustainability planning represents the next stage in strategic planning, and some cities around the globe have adopted the unified approach. Understandably some organizations may continue to utilize strategic plans with elements of sustainability and resilience or separate sustainability and resilience plans. Organizations must reinforce plans with climate adaptation, climate mitigation strategies, and climate preparedness actions to achieve resilience. The path to resilience and sustainability for municipal governments, cities, counties, townships, and other organizations require an assured interaction of organizational dynamics of leadership, financial commitment, and successful implementation of policies and programs. The subsequent is a non-exhaustive list of dynamic components required for effective resilience and sustainability policies, programs, and projects.

Institutionalization and Organizational Commitment

Organizations commit funds to meet the resilience and sustainability goals and practical application and implementation of resilience and sustainability-related activities. Connecting resilience and sustainability initiatives directly to the budget process enable organizations to benefit from system-wide strategic planning. Organizations institutionalize such efforts with the positive reinforcements of resilience and sustainability ideals and organizational goals, establishing resilience and sustainability offices. The institutionalization of resilience and sustainability ensures durability and survivability during and after the leadership changes.

Administrative, Managerial, and Political Leadership

With financial commitment, capable and willing leadership is fundamental to successful resilience and sustainable program implementation. An exemplary leadership at all levels of organizations, managerial, administrative, and political, leads to resilient and sustainable organizations. Elected and appointed officials and staff share ideas and provide feedback to each other. Leaders enable and encourage the development and implementation of initiatives at all levels of organizations. The leadership role is evaluated within the framework of a broader governance perspective. Incorporating resilience and sustainability core values into its planning process gives an organization a defined long-term planning perspective and leadership improvement.

Measuring, Tracking, Observing, and Scrutinizing Data and Results

Measuring, tracking, observing, and scrutinizing resilience and sustainability initiatives and outcomes and corresponding impacts facilitate organizational improvements. The resilience outcomes tracked and measured in the context of budgetary implications enable better management of resources, accountability, and transparency of programs and policies executions. The strategic resilience and sustainability plans do not guarantee increased efficiencies unless designed to include measurements, tracking processes, and observation and scrutiny of data and outcomes, coupled with reporting mechanisms.

Reporting Mechanism

Organizations make the interim and final outcome results public and incorporate resilience and sustainability targets into budgetary and fiscal plans. By presenting outcomes from resilience and sustainability programs, organizations communicate improvements, increase transparency, and share the plans for the future.

Stakeholder Engagement

The internal and external stakeholders are more supportive and engaging when included in program and policy implementation and enhancement of service delivery and initiatives and programs to improve organizational efficiency and the quality of life in communities. The uniqueness and identity of the communities and organizations are recognized in the engagement processes.

Partnerships and Collaborations

Prosperous programs and initiatives involve inventive and pragmatic alliances to leverage resources in the world fraught with financial instabilities. Financing resilience and sustainability-related projects require the banking sector's engagement, higher education institutions, non-government agencies, and all factors of the private and public sectors. One of the fundamental characteristics in adopting resilience and sustainability in a strategic plan is an initial activity from planning to implementation. The partnerships between local governments, communities, academic institutions and private sectors facilitate organizations' strategic resilience and sustainability plans.

The processes of embedding resilience and sustainability in organizations and communities are arduous and fraught with obstacles. It requires resources, commitment, institutional and leadership support, and partnerships and collaborations. Each segment of creating a resilient and sustainable community is critical. Using the Quadruple Bottom Line planning and implementing resilience and sustainability-related initiatives results in positive societal and governance outcomes, with equitable, healthy, and economically and environmentally resilient communities.

Outcomes, Discussions, and Further Considerations

- Assess the implementation processes for resilience and sustainability strategies, initiatives, programs, and projects.
- Analyze the effectiveness of climate-resilient strategies in organizations.
- Discuss the effects of strategic resilience and sustainability plan.
- Evaluate the implementation strategies and Quadruple Bottom Line approach to project implementation.
- Advance the deliberations on the risks associated with climate change and pandemics.

References

Alibašić H (2017) Measuring the sustainability impact in local governments using the Quadruple Bottom Line. Int J Sustain Policy Pract 13(3):37–45

Alibašić H (2018a) Paris Climate Accord sans the USA: the role of local governments in confronting the effects of climate change and increasing resilience. Glob Stud J 11(2):33–39. https://doi.org/10.18848/1835-4432/CGP/v11i02/33-39

Alibašić H (2018b) Leadership strategies for embedding sustainability and resilience in organizations with an emphasis on sustainable energy. Creighton J Interdiscip Leadersh 4(1):17–26

Alibašić H (2018c) The nexus of sustainability and climate resilience planning: embedding climate resilience policies in local governments. Int J Clim Change Impacts Resp 10(2):21–33. https://doi.org/10.18848/1835-7156/CGP/v10i02/21-33

Alibašić H (2018d) Sustainability and resilience planning for local governments: the quadruple bottom line strategy. Springer. https://www.springer.com/us/book/9783319725673

Alibašić H (2020) The administrative and ethical considerations of climate resilience: the politics and consequences of climate change. Public Integr. https://doi.org/10.1080/10999922.2020.1838142

Alibašić H, Crawley W (2018) Developing and continuing sustainability-related academic programming: observations of emerging practices. Creighton J Interdiscip Leadersh 4(1):27–34

Alibašić H, Crawley W (2020) Applying good governance through policy labs: sustainable solutions for universities and local governments. Int J Soc Sustain Econ Soc Cult Context 17(1):13–25. https://doi.org/10.18848/2325-1115/CGP/v17i01/13-25

Austin JR (1997) A method for facilitating controversial social change in organizations: Branch Rickey and the Brooklyn Dodgers. J Appl Behav Sci 33:101–118

Carbonn Center.Org (2021) The carbonn® Center & map of reporting local and regional governments as of July 30, 2021. https://carbonn.org/

Cities Alliance (n.d.) New joint work programme aims to help cities respond to climate change. https://www.citiesalliance.org/newsroom/news/urban-news/new-joint-work-programme-aims-help-cities-respond-climate-change

City of Golden Valley (2021) Resilience & sustainability. https://www.goldenvalleymn.gov/environmental/index.php

City of Golden Valley (n.d.) 2040 Comprehensive plan. Chapter 7: Resilience and sustainability. https://www.goldenvalleymn.gov/planning/comprehensiveplanupdate/pdf/compplanchapters/7-Resilience-and-Sustainability.pdf

Greenhouse Gas Protocol (n.d.) Global protocol for community-scale greenhouse gas emission inventories: an accounting and reporting standard for cities. https://ghgprotocol.org/sites/default/files/standards/GHGP_GPC_0.pdf

Johns Hopkins University & Medicine (2021) Mortality in the most affected countries. https://coronavirus.jhu.edu/data/mortality

Johns Hopkins University & Medicine (2022) Coronavirus resource center. https://coronavirus.jhu.edu/

Lorenz EN (1970) Climate change as a mathematical problem. J Appl Meteorol 9:325–322

Osborne D, Plastrik P (1997) Banishing bureaucracy: the five strategies for reinventing government. Addison-Wesley, Reading

Reckien D, Salvia M, Heidrich O, Church JM, Pietrapertosa F et al (2018) How are cities planning to respond to climate change? Assessment of local climate plans from 885 cities in the EU-28. J Clean Prod 191:207–219. https://doi.org/10.1016/j.jclepro.2018.03.220

Salvia M, Reckien D, Pietrapertosa F, Eckersley P, Spyridaki N, Krook-Riekkola A et al (2021) Will climate mitigation ambitions lead to carbon neutrality? An analysis of the local-level plans of 327 cities in the EU. Renew Sustain Energy Rev 135:110253. https://doi.org/10.1016/j.rser.2020.110253

Satterhwaite D (2008) Cities' contribution to global warming: notes on the allocation of greenhouse gas emissions. Environ Urban 20(2):539–549. https://doi.org/10.1177/0956247808096127

United Nations (2015) Transforming our world: the 2030 agenda for sustainable development A/Res/70/1. United Nations, New York. https://sustainabledevelopment.un.org/content/documents/21252030%20Agenda%20for%20Sustainable%20Development%20web.pdf

United Nations (n.d.) Climate action. https://www.un.org/en/climatechange/climate-solutions/cities-pollution

United Nations Department of Economic and Social Affairs Population Division (2018) The world's cities in 2018—data booklet (ST/ESA/ SER.A/417). https://www.un.org/en/events/citiesday/assets/pdf/the_worlds_cities_in_2018_data_booklet.pdf

United Nations Environment Programme (2020) UNEP emissions gap report 2020. United Nations Environment Programme. https://www.unep.org/emissions-gap-report-2020

United Nations General Assembly (UNGA) (1987) Report of the World Commission on Environment and Development: our common future. Transmitted to the General Assembly as an Annex to document A/42/427 - Development and International Co-operation: Environment. http://www.un-documents.net/ocf-02.htm

Washington County (n.d.) Washington County 2040 comprehensive plan: Chapter 9 resilience and sustainability. https://www.co.washington.mn.us/DocumentCenter/View/18620/Resilience-and-Sustainability?bidId

Wei T, Wu J, Chen S (2021) Keeping track of greenhouse gas emission reduction progress and targets in 167 cities worldwide. Front Sustain Cities. https://doi.org/10.3389/frsc.2021.696381

Weick KE (1993) Organizational redesign as improvisation. In: Huber GP, Glick WH (eds) Organizational change and redesign. Oxford University Press, New York, pp 346–379

Weick KE (1995) Sensemaking in organizations. Sage, Thousand Oaks

Weick KE, Quinn RE (1999) Organizational change and development. Annu Rev Psychol 50:361–386

Wiedmann T, Chen G, Owen A, Lenzen M, Doust M, Barrett J et al (2020) Three-scope carbon emission inventories of global cities. J Ind Ecol 25:735–750. https://doi.org/10.1111/jiec.13063

Index

H. Alibašić, *Strategic Resilience and Sustainability Planning*, Sustainable Development Goals Series, https://doi.org/10.1007/978-3-030-91159-1

The manufacturer's authorised representative in the EU is Springer Nature Customer Service Centre GmbH, Europaplatz 3, 69115 Heidelberg, Germany. If you have any concerns regarding our products, please contact ProductSafety@springernature.com

Printed and bound by CPI Group (UK) Ltd, Croydon, CR0 4YY
30/09/2026
02225710-0012